mathematics on the commodore 64

essential routines for programming

czes kosniowski

First published 1983 by:
Sunshine Books (an imprint of Scot Press Ltd.)
12–13 Little Newport Street,
London WC2R 3LD
01-437-4343

Copyright © Czes Kosniowski
Reprinted 1983, April 1984

ISBN 0 946408 14 9

All rights reserved. No part of this publication may be reproduced, stored in a retrieval system, or transmitted in any form or by any means, electronic, mechanical, photocopying, recording and/or otherwise, without the prior written permission of the Publishers.

Cover design by Grad Graphic Design Ltd.
Illustration by Stuart Hughes.
Typeset and printed in England by Commercial Colour Press, London E7.

CONTENTS

		Page
Preface		7
1	Simple functions?	9
2	Trigonometry	15
3	Earth trigonometry	33
4	Powers	37
5	Sequences	55
6	Number Bases	65
7	Days and Weeks	75
8	Greatest Common Divisor	85
9	Primes	89
10	Odds and Ends	99
11	Matrices	105
12	Codes	119
13	Random!	131
14	Meaningful Data	141
Summary		151

Contents in detail

CHAPTER 1
Simple Functions
Displaying numbers neatly, rounding off numbers, bank balances, overdrawn bank balances, colourful balances.

CHAPTER 2
Trigonometry
Scale drawings, the trigonometry functions, inverse functions, non right-angled triangles, refraction, reflection.

CHAPTER 3
Earth Trigonometry
The Earth, lines of longitude and latitude, calculating distances.

CHAPTER 4
Powers
Square roots, imaginary numbers, quadratic equations, solving other equations, Newton's method, exponential functions, logarithmic functions, roots of other functions.

CHAPTER 5
Sequences
Arithmetic sequences, geometric sequences, calculating interest, double or quit, Fibonacci sequences.

CHAPTER 6
Number Bases
Decimal representation, coefficients, binary numbers, hexadecimals, base converter, 64 numbers, small numbers, floating points.

CHAPTER 7
Days and Weeks
Calculating dates — Zeller's formula, calendar, date management.

CHAPTER 8
Greatest Common Divisor
Common factors, greatest common divisor, the Euclidean algorithm.

CHAPTER 9
Primes
Prime and composite numbers, testing primes, Sieve of Erastosthenes, large primes, Mersenne numbers, probabilistic primality testing, pseudoprimes.

CHAPTER 10
Odds and Ends
Pythagorean triplets and multi-precision powers.

CHAPTER 11
Matrices
Introducing matrices, adding matrices and how it can help, matrix multiplication, and why, zero matrices, identity matrices, inverses of matrices, simultaneous equations.

CHAPTER 12
Codes
Substitution codes, matrix codes, public key codes, encoding and decoding messages.

CHAPTER 13
Random!
Heads and tails, of dice and men, playing cards, non-equally likely events.

CHAPTER 14
Meaningful Data
Handling large amounts of numerical data, the mean, max, min and spread, standard deviation and variance, confidence intervals.

Preface

This book is written for all those who own a Commodore 64 and would like to know that little bit more about some mathematical techniques. You probably know what program you want to write but maybe you are not quite sure of the mathematics needed. Is it COS, ABS, or SGN that you need?

All the mathematical functions that you find on the Commodore 64 are described and their use is illustrated in short programs. You can 'lift' these programs and utilize them within your own programs.

But this book is not just an introduction to these basic mathematical functions. It contains background information and programs on such diverse subjects as codes and cryptography, random numbers, sequences, trigonometry, prime numbers, and statistical data analysis. You can utilize this information in both serious and games programming.

Many thanks to Ann, Kora and Inga for bearing with me during the writing of this book.

Czes Kosniowski
Newcastle upon Tyne, September 1983

Program notes

The Commodore 64 uses 'control characters' to control features such as cursor movement and colour printing. These control characters usually appear as inverse characters. For instance a reverse heart within quotes would clear the screen and move the cursor to its home position. To avoid difficulties in program listings control characters have not been used. Instead, their CHR$ equivalents have been used. A list of the more commonly used ones is provided below. Others may be found in Appendix F of the Commodore 64 User Manual.

    CHR$(5)   White
    CHR$(17)   Cursor down
    CHR$(28)   Red
    CHR$(30)   Green
    CHR$(31)   Blue
    CHR$(147) Clear screen and cursor home
    CHR$(145) Cursor up
    CHR$(154) Light Blue
    CHR$(157) Cursor left
    CHR$(158) Yellow
    CHR$(159) Cyan

When typing out the programs you may prefer to use control characters instead of the CHR$ statements.

Note. The symbol ^ should be entered as ↑ on the Commodore 64.

Owing to the quality of printout from the microcomputer printer, in some of the listings semicolons look very like colons. Please be careful when entering programs that you key in the correct character.

CHAPTER 1
Simple Functions?

Displaying numbers neatly

A whole number or a number without any decimal part is called an integer. Displaying integers neatly on your screen may be achieved by using the following lines in a program.

L = LEN(STR$(X))
PRINT TAB(25-L) X

The function STR$(X) converts the number X into a string, LEN calculates its length and TAB moves the cursor to the appropriate position on the screen.

```
        9
      123
     - 10
       89
```

For non-integral numbers the display goes astray.

```
       89
      1.2
   - 13.89
     .126
```

The numbers are right-justified; but it would be nice to have the decimal points vertically aligned. This may be achieved by using the functions INT(X) and ABS(X).

The function INT(X) returns the integral part of X, that is, the largest integer which is less than or equal to X. For example

INT(1.21) = 1
INT(2) = 2
INT(2.1) = 2
INT(-2) = -2

9

```
INT(-2.1) = -3
INT(9.1) = 9
INT(-9.2) = -10
```

The function ABS(X) returns the absolute value of X, that is, the number ignoring the + or - sign. For instance

```
ABS(9.1) = 9.1
ABS(-9.1) = 9.1
```

The following program displays numbers neatly on your screen.

```
Y = INT(ABS(X)) : L = LEN(STR$(Y))
IF Y = 0 AND X <> 0 THEN L = L-1
PRINT TAB(25-L) X
```

For example, a typical display is shown below

```
        3
         .23
      -89.14
   6712399.1
        2.23871
       -1.22
        -.13
```

In the first line of the program ABS takes care of negative numbers while INT takes care of non-integral numbers. Note that the INT function alone does not achieve this (for example look at the number -9.1). The second line takes care of numbers in the range greater than -1 and less than 1.

The program above illustrates one simple use of the functions INT and ABS. It works except for numbers which are close to 0 (absolute value less than or equal to 0.01) or which are very large (absolute value greater or equal to 1000000000). In fact whenever the scientific notation E appears the display goes slightly astray.

```
        3
      -89.14
        1E-04
   9.9E+16
```

You might like to add two lines to our program to take care of numbers involving scientific notation E.

Chapter 1 Simple Functions

Rounding off numbers

The INTegral function is useful for 'rounding off' numbers. For instance if you had £565.58 in a bank account and received 9% interest per annum then the amount you expect to have after one year is

   565.58 + 565.58*9/100

Using your Commodore 64 you can check that this has a value of 616.4822. But, of course, the bank would 'round' this *down* to £616.48. Similarly an amount such as 76.6752 would be rounded *up* to £76.68. Your Commodore 64 can do this rounding off with the following line.

   X = INT(X*100 + 0.5)/100

Here X is first multiplied by 100 to convert to pence. Then 0.5 is added which causes a rounding up if the fraction of pence is greater or equal to one-half. The INT function ignores any decimal parts and finally dividing by 100 converts the number back to pounds.

In general, the program line

   B = INT(A*10↑D + 0.5)/10↑D

gives the value of A rounded off to D decimal places.

Bank balances

The program given earlier on in this chapter for displaying numbers neatly on the screen could be used, for example, in a bank balance program. Using it (whenever a number is displayed) you may end up with a display such as below.

DETAILS	PAYMENTS	RECEIPTS	BALANCE
B/F			596.61
869162	46.22		550.39
869164	169		381.39
869165	15.01		366.38
CHQS		75.7	442.08

Observe that when the amount includes zero pence or a multiple of ten pence then the zero is not shown. To force this to happen we need to convert our numbers into STRings and add the necessary number of zeros. The following program lines illustrate how this may be achieved.

   X$ = STR$(X) : L = LEN(STR$(X*100)) : M = LEN(X$)
   IF M = L THEN X$ = X$ + "0"
   IF M = L−2 THEN X$ = X$ + ".00"
   PRINT TAB(25−L) X$

Notice that we do not need to use the INT and ABS functions. The bank statement display given earlier on would now appear as follows.

DETAILS	PAYMENTS	RECEIPTS	BALANCE
B/F			596.61
869162	46.22		550.39
869164	169.00		381.39
869165	15.01		366.38
CHQS		75.70	442.08

Overdrawn bank balances

Bank balances occasionally become overdrawn (or go into the red). This occurs when the balance becomes negative (less than zero). Thus a balance of −£64.00 means that you are overdrawn by £64.00. The following program lines illustrate how the function ABS may be used to print bank balances and indicate when the amount shown is overdrawn.

```
X$ = STR$(ABS(X)) : L = LEN(STR$(X*100)) : M = LEN(X$)
IF M = L THEN X$ = X$ + "0"
IF M = L-2 THEN X$ = X$ + ".00"
PRINT TAB(25-L) X$;
IF X < 0 THEN PRINT " DR";
PRINT
```

For example:

DETAILS	PAYMENTS	RECEIPTS	BALANCE
B/F			442.08
869166	52.80		389.29
869167	422.00		32.72 DR

Colourful balances

The function SGN(X) is the sign function which returns the sign (positive, negative, or zero) of the number X. The result is +1 if the number is positive, −1 if it is negative, and 0 if it is zero. For example:

```
SGN(9.21) = 1
SGN(-9.1) = -1
SGN(0) = 0
```

A typical use of the SGN function is when the program is required to perform different subroutines depending upon whether the sign of a number is positive, negative, or zero. For example, the program line

```
ON SGN(X) + 2 GOSUB 1000, 1100, 1200
```

would cause the program to execute the subroutine 1000 if X is negative, subroutine 1100 if X is 0, and subroutine 1200 if X is positive.

Chapter 1 Simple Functions

An interesting use of the SGN function is a simple method of changing the colour of printing. On the Commodore 64, CHR$(28) represents red while CHR$(30) represents green. Thus CHR$(29+SGN(X)) will be red or green depending on whether X is negative or positive. The following program adds this colourful feature to our money displaying program. The first POKE is used to give a background colour which enables green and red to be easily visible.

```
POKE 53281,7
X$ = STR$(ABS(X)) : L = LEN(STR$(X*100)) : M = LEN(X$)
IF M = L THEN X$ = X$ + "0"
IF M = L-2 THEN X$ = X$ + ".00"
PRINT CHR$(29+SGN(X)) TAB(25-L) X$;
IF X < 0 THEN PRINT " DR";
PRINT CHR$(30)
```

This short program displays numbers with two decimal parts (for example pounds and pence) so that they are vertically aligned. In addition the ABSolute value of a negative number is printed in red with DR after it.

CHAPTER 2
Trigonometry

Scale drawings

Seldom can we directly measure the heights of tall buildings, hills, trees, etc. One way to find the height of a building or tree is to stand away from the object. Now measure the angle between the horizontal and the highest point of the object (using a clinometer, which is just a glorified protractor), then measure the distance between you and the object. By drawing a scale drawing the height of the object can be readily estimated. See **Figure 1**.

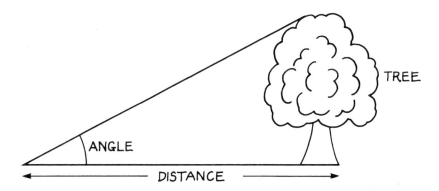

Figure 1.

You couldn't use the same technique to measure the height of a mountain peak which is miles away and covered in clouds. The clouds would get in your way, and you couldn't measure the horizontal distance. An instrument such as a tellinometer would help. This uses radar to locate the top of the mountain. It also measures the angle and distance between you and the top. A scale drawing would provide a way of calculating the height of the mountain. See **Figure 2**.

As a further example suppose we wanted to find the width of a large pond or lake; see **Figure 3**.

15

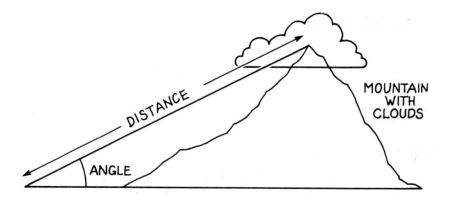

Figure 2.

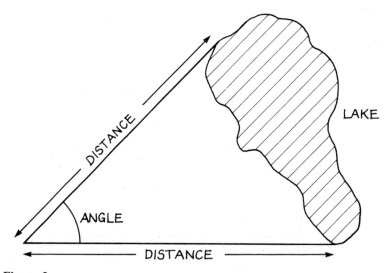

Figure 3.

A scale drawing drawn from the measurements made could be produced and the required distance estimated.

Here is a related example. A navigator is at a certain position A. He is 150 km due west of city B and 188 km from city C. The angle between the two cities is 23 degrees measured from his position. How far apart are the two cities? Again, a scale drawing could provide the answer.

Although scale drawing will provide answers to the problems mentioned above they are rough and ready. And it is not always practicable or accurate to produce scale drawings. An alternative approach is to do it by trigonometry using your Commodore 64.

Chapter 2 Trigonometry

The trigonometry functions

The three important trigonometric functions are SIN (sine function), COS (cosine function) and TAN (tangent function). They each represent ratios of the various sides of a right-angled triangle. For example, the triangle shown below is a right-angled triangle. The angle at the corner of the left is denoted by the symbol X. The three sides of the triangle will be referred to as the side adjacent to X, the side opposite X, and the hypotenuse (the longest side).

$$TAN(X) = \frac{opposite}{adjacent}$$

$$SIN(X) = \frac{opposite}{hypotenuse}$$

$$COS(X) = \frac{adjacent}{hypotenuse}$$

Some useful values to remember are the following:

SIN(0°) = 0	COS(0°) = 1	TAN(0°) = 0
SIN(30°) = 0.5	COS(30°) = SQR(3)/2	TAN(30°) = 1/SQR(3)
SIN(45°) = 1/SQR(2)	COS(45°) = 1/SQR(2)	TAN(45°) = 1
SIN(60°) = SQR(3)/2	COS(60°) = 1/2	TAN(60°) = SQR(3)
SIN(90°) = 1	COS(90°) = 0	

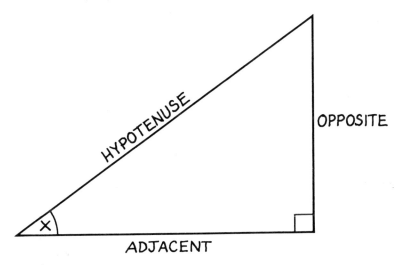

Figure 4.

If you know the angle X and one of the three lengths of a right-angled triangle then you can find the other two lengths. For example, if you know the angle X and the length of the adjacent side then the other two lengths are given by the following formulae.

opposite = TAN(X) * adjacent
hypotenuse = adjacent / COS(X)

Another way of describing the trigonometric function is by using a circle of radius 1 unit. Measure out the angle required as shown in **Figure 5**. The values of the various trigonometric functions are indicated.

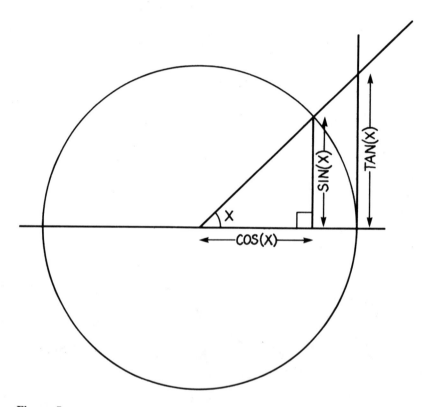

Figure 5.

Mathematically distances are measured horizontally from left to right and vertically upwards. This explains why, for instance, in the **Figure 6** COS(X) has a negative value.

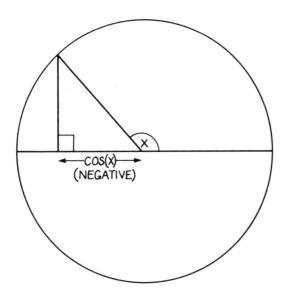

Figure 6.

You can obtain SIN, COS and TAN of an angle X by typing
    PRINT SIN(X) etc.,
substituting the appropriate value of X. The only possible problem is that the Commodore 64, like most microcomputers, expects the angles in radians, not degrees. Fortunately degrees can be turned into radians and vice versa very easily.

First of all, what is a radian? Draw a circle of radius 1 unit. Measure along the circumference of your circle a distance which is equal to the radius of the circle. The angle subtended by this arc is 1 radian. 1 radian is approximately 57°. See **Figure 7**.

The number π (or PI) is both remarkable and famous. It is defined to be the ratio of the circumference of a circle to its diameter. The (approximate) value of π is stored in your Commodore 64. Simply type
    PRINT π
to reveal the value stored. In a circle of radius 1 unit the diameter is 2 units. Thus the circumference of the circle is 2*π and so there are 2*π radians in a complete circle. Since there are 360 degrees in a complete circle we see that

    360 degrees = 2*π radians, and
    180 degrees = π radians

Mathematics on the Commodore 64

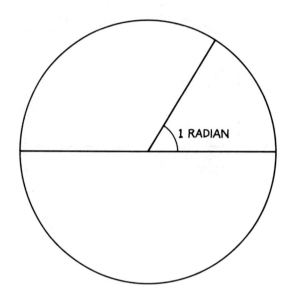

Figure 7.

We can convert degrees to radians and vice versa quite easily with the following formulae.

X degrees = X*π/180 radians
Y radians = Y*180/π degrees

The following program can be used to find lengths of right angled triangles. You need to input an angle and one distance. The program calculates the other two lengths.

```
10 REM PROGRAM FOR RIGHT-ANGLED TRIANGLES
20 PRINT CHR$(147) "        RIGHT-ANGLED TRIANGLES" CHR$(17)
30 PRINT "THIS PROGRAM ENABLES YOU TO FIND THE"
40 PRINT "SIDES OF A RIGHT ANGLED TRIANGLE"
50 PRINT "PROVIDED YOU KNOW ONE SIDE AND ANGLE." CHR$(17)
```

```
60 PRINT "                      **
                              *  *"
70 PRINT "                   *     *
                           *        *"
80 PRINT "      HYPOTENUSE*         *OPPOSITE
                        *            *"
90 PRINT "           *ANGLE *
                  **********"
95 PRINT "          ADJACENT" CHR$(17
)
100 REM INPUT DETAILS
110 INPUT "ANGLE, IN DEGREES, ";X
120 IF X<=0 OR X>=90 THEN PRINT "ERROR -
 NOT A TRIANGLE":GOTO 110
130 PRINT CHR$(17) "WHICH SIDE DO YOU KN
OW? 1 (OPPOSITE)"
140 PRINT "2 (ADJACENT) OR 3 (HYPOTENUSE
)."
150 INPUT "TYPE 1, 2 OR 3 ";T
160 IF T<1 OR T>3 OR T<>INT(T) THEN 150
170 PRINT CHR$(17) "TYPE IN THE LENGTH O
F THIS SIDE."
180 INPUT "LENGTH ";L
190 IF L<=0 THEN PRINT "FUNNY - TRY AGAI
N":GOTO 180
200 REM CONVERT TO RADIANS
210 X=X*π/180
220 REM THE SPLIT OFF
230 ON T GOSUB 300,350,400
240 PRINT CHR$(17) "     THAT'S IT - ANOT
HER GO Y OR N?"
250 GET G$:IF G$<>"Y" AND G$<>"N" THEN 2
50
260 IF G$="Y" THEN RUN
270 PRINT CHR$(147) "BYE FOR NOW.":END
300 REM OPPOSITE SIDE KNOWN
310 PRINT CHR$(17) "ADJACENT SIDE:" L/TA
N(X)
320 PRINT "   HYPOTENUSE:" L/SIN(X)
330 RETURN
350 REM ADJACENT SIDE KNOWN
360 PRINT CHR$(17) "OPPOSITE SIDE:" TAN(
X)*L
370 PRINT "   HYPOTENUSE:" L/COS(X)
```

```
380 RETURN
400 REM HYPOTENUSE KNOWN
410 PRINT CHR$(17) "OPPOSITE SIDE:" SIN(
X)*L
420 PRINT "ADJACENT SIDE:" L*COS(X)
430 RETURN

READY.
```

Inverse functions

Suppose we know the lengths of the sides of a right-angled triangle, can we determine the various angles? The answer is yes, and we use the inverse trigonometric functions to do this. Given an angle X then TAN(X) gives us a number, the tangent of the angle X. Conversely, given a number N we could find an angle whose tangent is that number. Such an angle could then be called the inverse tangent of N. It is usually denoted by ATN(N), the arc tangent of N.

Look at the triangle in **Figure 8**.

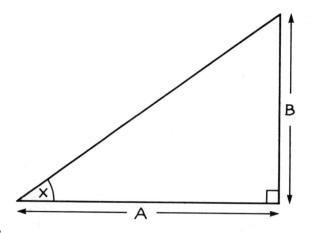

Figure 8.

If we know the values of A and B then we could find the value of the angle X. We know that TAN(X) = B/A, thus X = ATN(B/A). You can put in the appropriate values in this expression and get your Commodore 64 to print out the answer. Of course, the answer would be in radians. To get an answer in degrees you need to multiply the result by 180/π.

The trigonometric functions SIN and COS also have inverse functions denoted by ASN (arc sine) and ACS (arc cosine) respectively. ASN(N) is

Chapter 2 Trigonometry

that angle whose sine is N; similarly ACS(N) is that angle whose cosine is N. Unfortunately the Commodore 64, in common with many other microcomputers, does not contain these functions in its basic. However they can be easily obtained from the ATN function.

To see how we obtain ASN from ATN look at the right-angled triangle with a hypotenuse of length 1 unit in **Figure 9**.

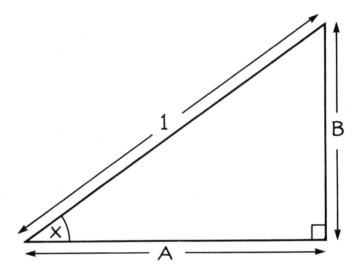

Figure 9.

Now suppose that we know the value of B, and we want to find the angle X. We know that SIN(X) = B so that X = ASN(B), but as was mentioned ASN isn't present in the Commodore 64. If we knew the value of A then we could use ATN since X = ATN(B/A) also. To find A we use Pythagoras' theorem.

Recall the theorem of Pythagoras. In words Pythagoras' theorem states that the square of the hypotenuse of a right-angled triangle is equal to the sum of the squares of the other two sides. In symbols we have

$$C^2 = A^2 + B^2$$

where C is the length of the hypotenuse. Since our hypotenuse is of length 1 we have:

1 = A*A + B*B

or

A*A = 1 − B*B

and so

$$A = SQR(1 - B*B)$$

Since $X = ATN(B/A)$ we obtain

$$X = ATN(B/SQR(1 - B*B))$$

also, $ASN(B) = X$, and so we obtain

$$ASN(B) = ATN(B/SQR(1 - B*B))$$

In a similar way we could produce a formula for $ACS(A)$, one such is given below:

$$ACS(A) = \pi/2 - ATN(A/SQR(1 - A*A))$$

You should notice that $ACS(A) = \pi/2 - ASN(A)$. Appendix H of the Commodore 64 User Manual contains other examples of mathematical functions which may be useful but which are not part of the 64's basic.

Non right-angled triangles

The first two examples from the scale drawing section may be solved by using the Right-Angled Triangles program. The third example (usually) involves non right-angled triangles.

A triangle has three angles and three sides. If we know the values of any three of these (except three angles) then we can find the values of the other three. For example we might know the length of two sides and one angle. We can then find the length of the third side and the value of the other two angles. To do this we use a formula.

Let's call the three angles in our triangle X, Y and Z; the three sides SX, SY and SZ where side SX is opposite angle X, etc. See **Figure 10**.

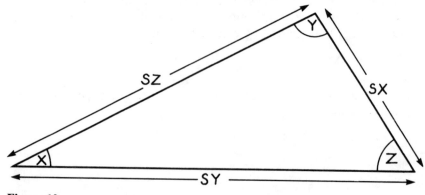

Figure 10.

Chapter 2 Trigonometry

The following formulae relate the various sides and angles.

The law of cosines:

$SZ*SZ = SX*SX + SY*SY - 2*SX*SY*COS(Z)$

$SY*SY = SX*SX + SZ*SZ - 2*SX*SZ*COS(Y)$

$SX*SX = SY*SY + SZ*SZ - 2*SY*SZ*COS(X)$

The law of sines: $SIN(X)/SX = SIN(Y)/SY = SIN(Z)/SZ$

Notice that if Z is a right-angle (that is 90 degrees) then $COS(Z) = 0$ and so the first formula becomes:

$SZ*SZ = SX*SX + SY*SY$

which is just Pythagoras' theorem.

The next program will find the remaining angles and sides provided you know one of the following:

Side Side Side : You know all three sides and are looking for the measurements of the three angles.

Side Side Angle : You know two sides and an angle which is not between them (a non-inclusive angle) and you are looking for the other side and angles.

Side Angle Side : You know two sides and the angle between them (the inclusive angle) and you are looking for the other side and angles.

Side Angle Angle : You know two angles and a side which is not between them (a non-inclusive side) and you are looking for the other two sides and the third angle.

Angle Side Angle : You know two angles and the side between them (the inclusive side) and you are looking for the other two sides and the third angles.

Notice that in the second case (Side Side Angle) two different triangles are (usually) possible depending on whether the angle opposite side 3 is greater than or less than 90 degrees. See **Figure 11** which illustrates this point.

```
10 REM TRIANGLES
20 PRINT CHR$(147)."      TRIANGLES" CHR$(
17)
30 PRINT "THIS PROGRAM WILL FIND THE REM
AINING"
40 PRINT "SIDES AND ANGLES OF A TRIANGLE
." CHR$(17)
```

25

```
50 PRINT "WHICH INFORMATION DO YOU HAVE?
" CHR$(17)
60 PRINT "1> SSS : ALL 3 SIDES" CHR$(17)
70 PRINT "2> SSA : 2 SIDES AND NON-INCLU
SIVE ANGLE"
80 PRINT "3> SAS : 2 SIDES AND INCLUSIVE
 ANGLE" CHR$(17)
90 PRINT "4> SAA : 2 ANGLES AND NON-INCL
USIVE SIDE"
100 PRINT "5> ASA : 2 ANGLES AND INCLUSI
VE SIDE" CHR$(17)
110 REM MAKE SELECTION
120 INPUT "TYPE IN NUMBER ": N
130 IF N<1 OR N>5 OR N<>INT(N) THEN PRIN
T."TRY 1, 2, 3, 4 OR 5.":GOTO 120
140 REM DEFINE ARCSINE FUNCTION, IN DEGR
EES TO 2 DECIMAL PLACES
150 DEF FNAS(X) = INT(18000*ATN(X/SQR(1-
X*X))/π + .5)/100
160 REM SPLIT OFF
170 PRINT:ON N GOSUB 310,510,710,910,101
0
180 PRINT CHR$(17),"ANOTHER GO? Y OR N"
190 GET G$:IF G$<>"Y" AND G$<>"N" THEN 1
90
200 IF G$="Y" THEN RUN
210 PRINT CHR$(147) "BYE FOR NOW.":END
300 REM ALL 3 SIDES
310 PRINT "*** ALL 3 SIDES KNOWN ***" CH
R$(17)
320 M=1:GOSUB 1110:SX=S
330 M=2:GOSUB 1110:SY=S
340 M=3:GOSUB 1110:SZ=S
350 A=(SY*SY+SZ*SZ-SX*SX)/(2*SY*SZ)
360 IF ABS(A)>=1 THEN PRINT "NOT A TRIAN
GLE":RETURN
370 PRINT "ANGLE OPPOSITE SIDE 1 IS " 90
-FNAS(A)
380 A=(SX*SX+SZ*SZ-SY*SY)/(2*SX*SZ)
390 PRINT CHR$(17) "ANGLE OPPOSITE SIDE
2 IS " 90-FNAS(A)
400 A=(SX*SX+SY*SY-SZ*SZ)/(2*SX*SY)
410 PRINT CHR$(17) "ANGLE OPPOSITE SIDE
3 IS " 90-FNAS(A)
```

Chapter 2   Trigonometry

```
420 RETURN
500 REM 2 SIDES AND A NON-INCLUSIVE ANGL
E
510 PRINT "** 2 SIDES AND A NON-INCLUSIV
E ANGLE ** "
520 PRINT "TYPE IN THE SIDE FOR WHICH TH
E OPPOSITE ANGLE IS KNOWN" CHR$(17)
530 M=1:GOSUB 1110:SX=S:GOSUB 1210:AX=A
540 M=2:GOSUB 1110:SY=S
550 A=SIN(AX)*SY/SX:IF ABS(A)>1 OR A=0 T
HEN PRINT."NOT A TRIANGLE":RETURN
560 PRINT "IS ANGLE OPPOSITE SIDE 2 GREA
TER (>) OR LESS (<) THAN 90 DEGREES?"
570 INPUT "TYPE > OR < ": A$
580 IF A$<>"<" AND A$<>">" THEN 570
590 AY=FNAS(A):IF A$=">" AND AY<90 THEN
AY=90+AY
600 PRINT CHR$(17) "ANGLE OPPOSITE SIDE
2 IS" AY CHR$(17)
610 AZ=π-AX-AY*π/180
620 PRINT "LENGTH OF SIDE 3 IS" SX*SIN(A
Z)/SIN(AX) CHR$(17)
630 PRINT "ANGLE OPPOSITE SIDE 3 IS" INT
(18000*AZ/π + .5)/100
640 RETURN
700 REM 2 SIDES AND THE INCLUSIVE ANGLE
710 PRINT "*** 2 SIDES AND THE INCLUSIVE
 ANGLE *** "
720 M=1:GOSUB 1110:SX=S
730 M=2:GOSUB 1110:SY=S
740 M=3:GOSUB 1210:AZ=A
750 SZ=SQR(SX*SX+SY*SY-2*SX*SY*COS(AZ))
760 IF SZ=0 THEN PRINT."NOT A TRIANGLE":
RETURN
770 PRINT "LENGTH OF SIDE 3 IS" SZ CHR$(
17)
780 A=(SY*SY+SZ*SZ-SX*SX)/(2*SY*SZ)
790 PRINT "ANGLE OPPOSITE SIDE 1 IS " 90
-FNAS(A)
800 A=(SX*SX+SZ*SZ-SY*SY)/(2*SX*SZ)
810 PRINT CHR$(17) "ANGLE OPPOSITE SIDE
2 IS " 90-FNAS(A)
820 RETURN
900 REM 2 ANGLES AND A NON-INCLUSIVE SID
E
```

```
910 PRINT "** 2 ANGLES AND A NON-INCLUSI
VE SIDE ** "
920 PRINT "TYPE IN THE ANGLE FOR WHICH T
HE OPPOSITESIDE IS KNOWN FIRST" CHR$(17)
930 M=1:GOSUB 1210:AX=A:GOSUB 1110:SX=S
940 M=2:GOSUB 1210:AY=A
950 A=π-AX-AY:IF A<=0 THEN PRINT,"NOT A
 TRIANGLE":RETURN
960 PRINT "LENGTH OF SIDE 2 IS " SX*SIN(
AY)/SIN(AX) CHR$(17)
970 PRINT "ANGLE OPPOSITE SIDE 3 IS " IN
T(18000*A/π + .5)/100 CHR$(17)
980 PRINT "LENGTH OF SIDE 3 IS " SX*SIN(
A)/SIN(AX)
990 RETURN
1000 REM 2 ANGLES AND AN INCLUSIVE SIDE
1010 PRINT "*** 2 ANGLES AND AN INCLUSIV
E SIDE *** "
1020 M=1:GOSUB 1210:AX=A
1030 M=2:GOSUB 1210:AY=A
1040 M=3:GOSUB 1110:SZ=S
1050 A=π-AX-AY:IF A<=0 THEN PRINT,"NOT A
 TRIANGLE":RETURN
1060 PRINT "ANGLE OPPOSITE SIDE 3 IS " I
NT(18000*A/π + .5)/100 CHR$(17)
1070 PRINT "LENGTH OF SIDE 1 IS " SZ*SIN
(AX)/SIN(A) CHR$(17)
1080 PRINT "LENGTH OF SIDE 2 IS " SZ*SIN
(AY)/SIN(A)
1090 RETURN
1100 REM GET A SIDE
1110 S=0:PRINT "TYPE LENGTH OF SIDE" M "
: ";:INPUT S:PRINT CHR$(17);
1120 IF S<=0 THEN PRINT,"NOT A TRIANGLE!
":GOTO 1110
1130 RETURN
1200 REM GET AN ANGLE
1210 A=0:PRINT "TYPE ANGLE OPPOSITE SIDE
 " M ": ";:INPUT A:PRINT CHR$(17);
1220 IF A<=0.001 OR A>=180 THEN PRINT,"N
OT A TRIANGLE!":GOTO 1210
1230 A=A*π/180:RETURN

READY.
```

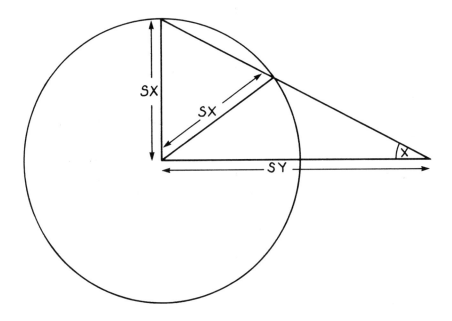

Figure 11.

Refraction

Things often look distorted when viewed through glass or plastic. Water looks shallower than it actually is. The reason is refraction. When a ray of light travels from one medium (air) to another (glass, water, ...) it is bent or refracted. The angle that the ray hits the glass with is called the angle of incidence; the angle after it has been refracted is called the angle of refraction — see **Figure 12**.

For a given material there is a fixed relation between the angles of incidence and refraction. This is given by Snell's law which states that the ratio of the sine is constant for any material (in air). This ratio is called the refractive index.

$$\text{refractive index} = \frac{\text{SIN(angle of incidence)}}{\text{SIN(angle of refraction)}}$$

For glass the refractive index is about 1.5, for water it is 1.333, while for diamond it is 2.417.

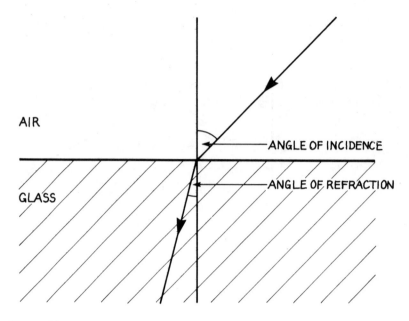

Figure 12.

The program allows you to determine the angle of refraction, assuming that you know the angle of incidence and the refractive index.

```
10 REM REFRACTION PROGRAM
20 PRINT CHR$(147) "             REFRACT
ION" CHR$(17)
30 PRINT "THIS PROGRAM CALCULATES THE AN
GLE OF"
40 PRINT "REFRACTION WHEN A RAY OF LIGHT
 HITS"
50 PRINT "ANOTHER MEDIUM." CHR$(17)
100 REM INPUT DETAILS
110 PRINT "TYPE IN ANGLE OF INCIDENCE. I
N DEGREES."
120 INPUT "ANGLE: ";X
130 IF X<=0 OR X>=90 THEN PRINT "ERROR -
 NONSENSE":GOTO 120
140 PRINT CHR$(17) "WHAT IS THE REFRACTI
VE INDEX OF THE    MEDIUM?"
145 INPUT "REFRACTIVE INDEX: ";R
150 IF R<=0 THEN PRINT "FUNNY - TRY AGAI
```

```
N":GOTO 145
160 REM CONVERT TO RADIANS
170 X=X*π/180
180 REM CALCULATE
190 Y=SIN(X)/R:Y=Y/SQR(1-Y*Y)
200 PRINT CHR$(17) "ANGLE OF REFRACTION:
 " ATN(Y)*180/π "DEGREES."
210 PRINT "PERCENTAGE OF ANGLE OF INCIDE
NCE:" INT(ATN(Y)*100/X)
240 PRINT CHR$(17) "    THAT'S IT - ANOT
HER GO Y OR N?"
250 GET G$:IF G$<>"Y" AND G$<>"N" THEN 2
50
260 IF G$="Y" THEN RUN
270 PRINT CHR$(147) "BYE FOR NOW.":END

READY.
```

Reflection

A piece of glass or the surface of water occasionally behaves like an ordinary mirror, reflecting everything. This occurs when the angle of incidence is too great and the ray of light is reflected. The smallest angle at which this occurs is called the critical angle of the medium. This is given by the following simple formula:

$$\text{SIN(critical angle)} = \frac{1}{\text{refractive index}}$$

Thus the critical angle can be determined from the refractive index by using the ASN function described earlier on.

CHAPTER 3
Earth Trigonometry

The Earth

The earth is almost a sphere. But it is slightly squashed at the north and south pole. The equator has a radius of 6378 kilometres. The polar radius is 6357 kilometres. The difference is only about 0.3% and you can hardly tell that it isn't a sphere. The average radius is 6371 kilometres.

We are used to saying that the shortest distance between two points is a 'straight line'. But this applies only on the plane, or on any 'flat' surface. On a sphere, like the Earth, the shortest distance between two points is part of a circle called a great circle. A great circle is a circle whose centre is the centre of the earth.

Great circles passing through the north and south poles are called *lines of longitude*. Lines of longitude have an angle associated to them. The line of longitude that passes through Greenwich, England is marked 0° (0 degrees). The others are marked by measuring the angle at the centre of the Earth between the line of longitude and the one at Greenwich — see **Figure 13**.

Usually longitude lines go from 0° to 180° both east and west.

Lines of longitude tell us how far a point on the Earth is east or west of Greenwich. To show how far a point is north or south of the equator we use lines of latitude. The equator is said to be of latitude 0°. Circles on the Earth which are parallel to the equator are called lines of latitude. The angle (measured at the centre of the Earth) between the equator and a line of latitude is called the latitude — see **Figure 14**.

Latitudes go from 0° to 90° both north and south. The north pole is at latitude 90° north, while the south pole is at 90° south.

Any point on the Earth may be pin-pointed by its latitude and longitude. For instance, Newcastle upon Tyne (England) is at latitude 55° north and 1.5° west approximately. More accurate values are 54° 58' N and 1° 36' W, where the symbol ' is read as minutes and a minute is 1/60 of a degree.

Latitudes and longitudes are a set of co-ordinates on the surface of the Earth.

Calculating the (shortest, great circle) distance between two points on the Earth is not easy. For instance, what is the distance between

Mathematics on the Commodore 64

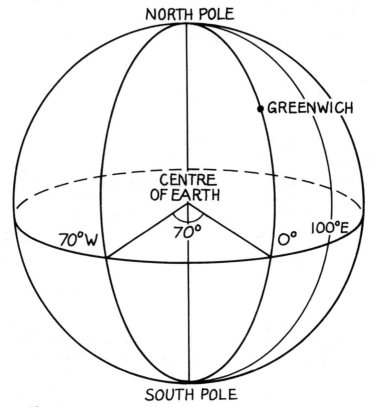

Figure 13.

Newcastle upon Tyne and Paris, France? (Paris is at approximately 49° N, 2° E.) With a suitable program on the Commodore 64 this presents no problems.

The next program calculates distances between two points on the Earth's surface. The mathematics behind the program is based on several uses of the cosine and sine rules discussed in the previous chapter.

```
10 REM EARTH TRIGONOMETRY
20 POKE 53281,6:PRINT CHR$(147),CHR$(154
   ) "EARTH TRIGONOMETRY" CHR$(17)
30 PRINT "THIS PROGRAM CALCULATES THE SH
   ORTEST"
40 PRINT "DISTANCE BETWEEN TWO POINTS ON
    THE EARTH "
50 REM INPUT DATA
60 FOR I=1 TO 2
70 PRINT:PRINT CHR$(5) "POSITION";I;CHR$
```

Chapter 3 Earth Trigonometry

```
(153) CHR$(17)
80 INPUT " LATITUDE ";A(I)
90 IF A(I)<0 OR A(I)>90 THEN PRINT "BETW
EEN 0 AND 90!!":GOTO 80
100 INPUT "    N OR S ";A$
110 IF A$<>"N" AND A$<>"S" THEN PRINT "N
ORTH OR SOUTH!!":GOTO 100
120 IF A$="S" THEN A(I)=-A(I)
130 PRINT:INPUT "LONGITUDE ";B(I)
140 IF B(I)<0 OR B(I)>180 THEN PRINT "BE
TWEEN 0 AND 180!!":GOTO 130
150 INPUT "   E OR W ";A$
160 IF A$<>"E" AND A$<>"W" THEN PRINT "E
AST OR WEST!!":GOTO 150
170 IF A$="E" THEN B(I)=-B(I)
180 NEXT
190 PRINT:PRINT CHR$(152) "DO YOU WANT T
HE DISTANCE IN MILES OR    KILOMETRES?"
200 INPUT "M OR K":A$:PRINT
210 IF A$<>"M" AND A$<>"K" THEN 200
220 R=6371:B$="KILOMETRES":IF A$="M" THE
N R=3960:B$="MILES"
230 REM THE CALCULATION
240 A1=A(1)*π/180:A2=A(2)*π/180:B1=B(1)*
π/180:B2=B(2)*π/180
250 B=ABS(B(1)-B(2)):IF B>180 THEN B=180
-B
260 A=ABS(A(1)-A(2))*π/360:B=B*π/360
270 X=COS(A1)*SIN(B)*COS(A2)*SIN(B)+SIN(
A)*SIN(A)
280 D=2*R*ATN(SQR(X/(1-X)))
290 PRINT CHR$(158) "THE DISTANCE IS";IN
T(D*100)/100;B$
300 PRINT:PRINT CHR$(154),"ANOTHER GO? Y
 OR N"
310 GET G$:IF G$<>"Y" AND G$<>"N" THEN 3
10
320 IF G$="Y" THEN RUN
330 PRINT CHR$(147) "BYE FOR NOW":END

READY.
```

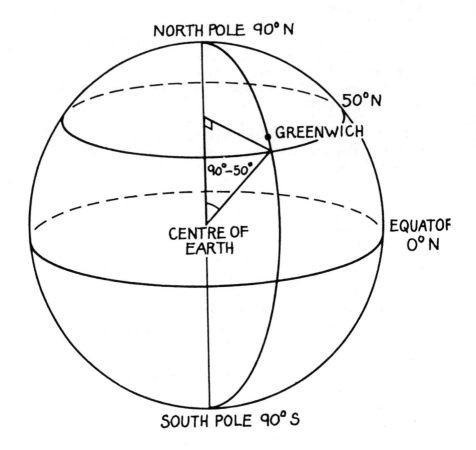

Figure 14.

CHAPTER 4
Powers

The square of a number X is the number multiplied by itself, that is X*X. It is denoted by X^2 or X↑2. The square of a number is also called the second power of X.

The powers of X are products of the appropriate number of X's. The first, second, third, fourth and fifth powers of the number 2 are the numbers

2, 4, 8, 16, and 32.

We denote these powers respectively by

2^1, 2^2, 2^3, 2^4, and 2^5.

You can calculate the powers of 2 (or any other number) with the following simple program.

```
10 REM POWERS
20 PRINT CHR$(147),"         POWERS" CHR$(17)
30 PRINT "THIS PROGRAM PRINTS OUT THE PO
WERS OF A NUMBER ENTERED." CHR$(17)
40 INPUT "NUMBER ";X:PRINT
50 INPUT "HOW MANY POWERS ";N:PRINT
60 IF N<1 OR N<>INT(N) THEN PRINT"AN INT
EGER PLEASE":GOTO 50
70 REM THE CALCULATION AND PRINT OUT
80 PRINT CHR$(5) "  I     X^I" CHR$(17)
90 Y=1
100 FOR I=1 TO N:Y=Y*X:PRINT I;Y:NEXT
110 PRINT:PRINT CHR$(154),"ANOTHER GO? Y
 OR N"
120 GET G$: IF G$<>"Y" AND G$<>"N" THEN 1
20
130 IF G$="Y" THEN RUN
140 PRINT CHR$(147) "BYE FOR NOW":END

READY.
```

Mathematics on the Commodore 64

If the number N is too large you'll get an overflow error.

You can add a few lines to test for a possible overflow and escape.

```
10 REM POWERS PLUS
20 PRINT CHR$(147),CHR$(154) "        POWE
RS" CHR$(17)
30 PRINT "THIS PROGRAM PRINTS OUT THE PO
WERS OF A NUMBER ENTERED." CHR$(17)
40 INPUT "NUMBER ";X:PRINT
50 INPUT "HOW MANY POWERS ";N:PRINT
60 IF N<1 OR N<>INT(N) THEN PRINT"AN INT
EGER PLEASE":GOTO 50
65 IF (N+1)*LOG(ABS(X))>126*LOG(2) THEN
PRINT "YOU'LL PROBABLY OVERFLOW":PRINT
70 REM THE CALCULATION AND PRINT OUT
80 PRINT CHR$(5) "  I     X^I" CHR$(17)
90 Y=1
100 FOR I=1 TO N:Y=Y*X:PRINT I;Y:NEXT
110 PRINT:PRINT CHR$(154),"ANOTHER GO? Y
 OR N"
120 GET G$:IF G$<>"Y" AND G$<>"N" THEN 1
20
130 IF G$="Y" THEN RUN
140 PRINT CHR$(147) "BYE FOR NOW":END

READY.
```

The Commodore 64 can also calculate the power N of a number X by using PRINT X↑N.

```
10 REM POWERS AGAIN
20 PRINT CHR$(147),CHR$(154) "      POWERS
 AGAIN" CHR$(17)
30 PRINT "THIS PROGRAM PRINTS OUT THE PO
WERS OF A NUMBER ENTERED." CHR$(17)
40 INPUT "NUMBER ";X:PRINT
50 INPUT "HOW MANY POWERS ";N:PRINT
60 IF N<1 OR N<>INT(N) THEN PRINT"AN INT
EGER PLEASE":GOTO 50
70 IF (N+1)*LOG(ABS(X))>126*LOG(2) THEN
PRINT "YOU'LL PROBABLY OVERFLOW":PRINT
80 REM THE CALCULATION AND PRINT OUT
```

```
90 PRINT CHR$(5) " I    X^I" CHR$(17)
100 FOR I=1 TO N:PRINT I;X^I:NEXT
110 PRINT:PRINT CHR$(154),"ANOTHER GO? Y
    OR N"
120 GET G$: IF G$<>"Y" AND G$<>"N" THEN 1
20
130 IF G$="Y" THEN RUN
140 PRINT CHR$(147) "BYE FOR NOW":END
```

READY.

As you may know or can soon discover, N does not have to be a whole number.

But what does 2↑1.7 mean? It is reasonable that it should be a number between 2↑1 and 2↑2, that is, between 2 and 4. Indeed, the value of 2↑1.7 is about 3.24900959. An approximate way of finding the value of 2↑1.7 (without using your computer) is to use some graph paper. Plot the powers of 2 from the first power to the fifth power — see **Figure 15**. Next draw a smooth curve through these points — see **Figure 16**. The approximate value of 2↑1.7 can be read from this graph — see **Figure 17**.

Similar graphs could be drawn for powers of other numbers. Of course, you need not do this since your Commodore 64 will give the answer immediately.

Negative powers of numbers also make sense, these are simply defined by the following rule:

$X^{-N} = 1/X^N$.

Thus $2^{-2} = 1/2^2$ which is 1/4 or 0.25. By convention, the zero-th power of a number is 1.

Powers of a number behave in a nice way according to the following rule:

$X^M * X^N = X^{M+N}$.

So that, for instance,

$3^4 * 3^2 = 3^6$

and

$10^{-2} * 10^3 = 10^1$.

Square roots

For a number X the number $X^{1/2}$ has a special name — it is called the *square root* of X. The square root of a number is that number whose square is the number. Thus the square root of 9 is 3 and the square root of 2 is approximately 1.41421356 as you can readily check by multiplying this number with itself.

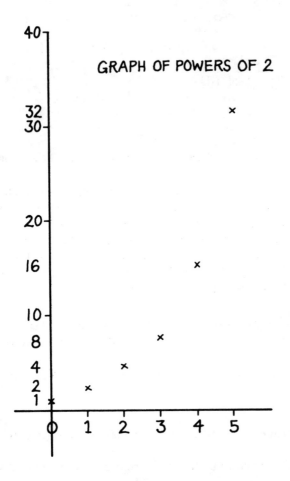

Figure 15

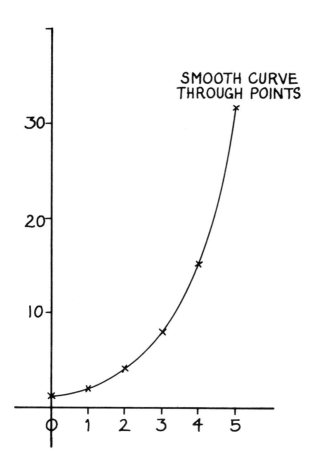

Figure 16

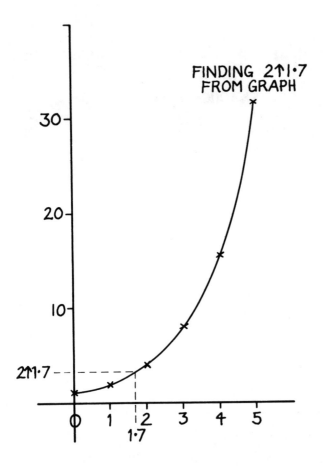

Figure 17

The notation $X^{1/2}$ for the square root of X fits into the way that multiplication of powers works:

$$X^{1/2} * X^{1/2} = X^1$$

which is X of course.

The square root of a number X can also be found with the Commodore 64 by using PRINT SQR(X).

Imaginary numbers?

The square of a number is always positive and so you should not expect to be able to find the square root of a negative number. Indeed if you ask the Commodore 64 to PRINT SQR(−1) it will respond with an ILLEGAL QUANTITY ERROR message.

You may have heard of imaginary numbers and complex numbers which are associated with square roots of negative numbers. Mathematicians are never deterred by seemingly impossible things such as the square root of −1. One simply creates a new symbol to stand for this number. Thus we let I stand for the square root of −1. There is nothing strange about this, I exists in the same way as negative numbers exist.

We can add the number I to itself, to other numbers and multiply it by other numbers. We can therefore form numbers such as

2 + 3*I, 1.41412*I, 9 − I, 10 − 8*I.

Numbers involving I are called complex numbers, usually to distinguish them from ordinary or real numbers. Any complex number may be written in the form

X + Y*I

for some real numbers X and Y. We call X the real part and Y the imaginary part of the complex number.

Once we decide to use the symbol I for SQR(−1) we can find square roots of other negative numbers.

$$\begin{aligned} SQR(X) &= SQR(ABS(X)*SGN(X)) \\ &= SQR(ABS(X))) * SQR(SGN(X)) \end{aligned}$$

Thus, for instance,

$$\begin{aligned} SQR(-9) &= SQR(9) * SQR(-1) \\ &= 3*I. \end{aligned}$$

But, of course, the Commodore 64 will not operate this way because it doesn't know about complex numbers.

Mathematics on the Commodore 64

The next program shows how you could incorporate complex numbers within your Commodore 64. Simple arithmetic operations may then be performed. Note that multiplication of complex numbers follows the following rule.

$$\begin{aligned}(A + B*I) * (C + D*I) &= A*C + A*D*I + B*I*C + B*I*D*I \\ &= A*C + A*D*I + B*C*I + B*D*I*I \\ &= A*C + A*D*I + B*C*I + B*D*-1 \\ &= A*C - B*D + (A*D + B*C)*I\end{aligned}$$

```
10 REM COMPLEX NUMBERS
20 PRINT CHR$(147),CHR$(154) " COMPLEX N
UMBERS" CHR$(17)
30 PRINT "THIS PROGRAM PERFORMS MULTIPLI
CATION"
40 PRINT "AND DIVISION OF COMPLEX NUMBER
S." CHR$(17)
50 PRINT "ENTER COMPLEX NUMBERS AS REQUE
STED." CHR$(17)
60 PRINT CHR$(158) "FIRST NUMBER" CHR$(1
7)
70 INPUT "      REAL PART ";A
80 INPUT "IMAGINARY PART ";B:PRINT
90 PRINT "SECOND NUMBER" CHR$(17)
100 INPUT "      REAL PART ";C
110 INPUT "IMAGINARY PART ";D:PRINT
120 PRINT CHR$(154) "WHAT DO YOU WANT TO
 DO?"
130 PRINT "1   MULTIPLY THE TWO NUMBERS."
140 PRINT "2   DIVIDE THE FIRST NUMBER BY
 THE SECOND " CHR$(158)
150 INPUT "ENTER 1 OR 2 ";N:PRINT
160 IF N<>1 AND N<>2 THEN 150
170 ON N GOSUB 300,350
180 PRINT:PRINT CHR$(154),"ANOTHER GO? Y
 OR N"
190 GET G$:IF G$<>"Y" AND G$<>"N" THEN 1
90
200 IF G$="Y" THEN RUN
210 PRINT CHR$(147) "BYE FOR NOW":END
300 REM MULTIPLICATION
310 PRINT CHR$(5) "THE PRODUCT OF THE NU
MBERS: "
320 PRINT "      REAL PART ";A*C-B*D
```

```
330 PRINT "IMAGINARY PART ":A*D+B*C
340 RETURN
350 REM DIVISION
360 X=C*C+D*D:IF X=0 THEN PRINT CHR$(5)
"DIVISION NOT POSSIBLE":RETURN
370 PRINT CHR$(5) "THE DIVISION OF THE F
IRST BY SECOND:"
380 PRINT "        REAL PART ":(A*C+B*D)/X
390 PRINT "IMAGINARY PART ":(-A*D+B*C)/X
400 RETURN
READY.
```

Quadratic equations

Quadratic equations often arise when solving problems of one sort or another. The general form of a quadratic equation is

$A*X^2 + B*X + C = 0$

where A, B and C are known numbers with A non-zero. The problem is to find those numbers X that satisfy the equation. These are called the *roots* of the quadratic equation. For example the values X = 1 and X = 2 satisfy the following quadratic equation

$X^2 - 3*X + 2 = 0$

as you can readily verify.

Usually there are two roots to a quadratic equation, often the roots are complex numbers.

There is a very straightforward formula which provides the roots of a quadratric equation. The roots of a quadratic equation are given by

$$\frac{-B + SQR(B^2 - 4*A*C)}{2*A}$$

and

$$\frac{-B - SQR(B^2 - 4*A*C)}{2*A}$$

The key to the nature of the roots is found in that part of the formula involving the square root function,

$SQR(B^2 - 4*A*C)$

which is called the *discriminant*.

If $B^2 - 4*A*C > 0$ then there are two real roots.
If $B^2 - 4*A*C = 0$ then the two roots are real and the same.
If $B^2 - 4*A*C < 0$ then there are two complex roots.

The next program calculates the roots of a quadratic equation.

```
10 REM QUADRATIC EQUATIONS
20 PRINT CHR$(147),CHR$(154) "QUADRATIC
EQUATIONS" CHR$(17)
30 PRINT "THIS PROGRAM SOLVES QUADRATIC
EQUATIONS"
40 PRINT "LIKE A*X*X + B*X + C = 0." CHR
$(17)
50 INPUT "VALUE OF A ";A:PRINT
60 IF A=0 THEN PRINT "A CANNOT BE ZERO.
TRY AGAIN.":GOTO 50
70 INPUT "VALUE OF B ";B:PRINT
80 INPUT "VALUE OF C ";C:PRINT
90 REM START OF CALCULATION
100 D=B*B-4*A*C
110 PRINT CHR$(158):ON SGN(D)+2 GOSUB 20
0,300,400
120 PRINT:PRINT CHR$(154),"ANOTHER GO? Y
 OR N"
130 GET G$:IF G$<>"Y" AND G$<>"N" THEN 1
30
140 IF G$="Y" THEN RUN
150 PRINT CHR$(147) "BYE FOR NOW":END
200 REM TWO COMPLEX ROOTS
210 PRINT "THERE ARE TWO COMPLEX ROOTS."
 CHR$(17)
220 X=-B/A/2:Y=ABS(SQR(-D)/A/2)
230 PRINT X;"+";Y;CHR$(157);"*I"
240 PRINT X;"-";Y;CHR$(157);"*I"
250 RETURN
300 REM TWO EQUAL ROOTS
310 PRINT "THE TWO ROOTS ARE EQUAL." CHR
$(17)
320 PRINT -B/A/2
330 RETURN
400 REM TWO REAL ROOTS
410 PRINT "THERE ARE TWO REAL ROOTS." CH
R$(17)
```

```
420 D=SQR(D):X=(-B-D)/2:IF B<0 THEN X=(-
B+D)/2
430 PRINT X/A:PRINT C/X
440 RETURN
```

READY.

If there are two real roots, then the program uses the quadratic formula to find one of the roots — the one with the largest absolute value. The other root is found by using the fact that the product of the two roots is C/A.

Solving other equations

Solving quadratic equations is relatively straightforward. This is not so for other equations such as

$$9*X^5 - 3*X^4 + X^3 - X^2 + 5*X - 4 = 0.$$

If the equation only involves non-negative integral powers of X, as in the example above, we call the equation a *polynomial* equation. The largest non-zero power of X that occurs is called the *degree* of the polynomial. In the above equation the degree is 5. The quadratic equation is a polynomial equation of degree 2. Quadratic equations usually have 2 roots; a polynomial equation of degree N usually has N roots.

Apart from some special cases there are no general formulae for solving polynomial equations. Indeed this lack has led to some very interesting mathematics, but that's another story and will not be covered here.

We can, however, use our Commodore 64 to work out roots of polynomial equations by a repeated guess-and-try process. Essentially the computer tries many different numbers to find out which satisfies the equation.

The next program provides a method of finding a real root of a polynomial equation. It is mathematically crude and occasionally will not find a root, even if there is one. The program does however illustrate the general method involved.

```
10 REM POLY ROOTS
20 PRINT CHR$(147),CHR$(154) "ROOTS OF P
OLYNOMIALS" CHR$(17)
30 PRINT "THIS PROGRAM ATTEMPTS TO FIND
ROOTS OF"
40 PRINT "POLYNOMIALS SUCH AS" CHR$(17)
50 PRINT "A*X^N + B*X^(N-1) + ... + C*X
+ D = 0."
```

```
60 INPUT "DEGREE OF POLYNOMIAL ";N:PRINT
70 IF N<2 OR N<>INT(N) THEN PRINT "INTEG
ER PLEASE.":GOTO 60
80 DIM A(N):PRINT "ENTER COEFFICIENTS TE
RM BY TERM." CHR$(17)
90 FOR I=0 TO N
100 IF I<N THEN PRINT "   COEFFICIENT OF
 X^";MID$(STR$(N-I),2);
110 IF I=N THEN PRINT "CONSTANT COEFFICI
ENT";
120 INPUT A(I)
130 IF A(0)=0 THEN PRINT "NOT ZERO PLEAS
E.":GOTO 100
140 NEXT
150 PRINT:PRINT "ENTER RANGE OVER WHICH
SEARCH OF ROOTS"
160 PRINT "IS TO BE ATTEMPTED." CHR$(17)
170 INPUT "LOWER VALUE ";A
180 INPUT "UPPER VALUE ";B:PRINT
190 IF A>=B THEN PRINT "FIRST VALUE SHOU
LD BE LOWER.":GOTO 170
200 REM SEARCH
210 S=B-A:T=0:TEST=-1:D=1E-9
220 PRINT "TEST RUN";T+1;"**";10^T;"DIVI
SIONS **":GOSUB 300
230 IF TEST THEN S=S/10:T=T+1:IF T<4 THE
N GOTO 220
240 IF T=4 THEN PRINT:PRINT "CANNOT LOCA
TE ROOTS - SORRY."
250 PRINT:PRINT CHR$(154),"ANOTHER GO? Y
 OR N"
260 GET G$:IF G$<>"Y" AND G$<>"N" THEN 2
60
270 IF G$="Y" THEN RUN
280 PRINT CHR$(147) "BYE FOR NOW":END
300 REM STEP BY STEP SEARCH
310 X=A:GOSUB 400:X1=A:Y1=Y
320 FOR X=A+S TO B STEP S
330 GOSUB 400:X2=X:Y2=Y
340 IF Y1*Y2<=D THEN GOSUB 500
350 Y1=Y2:X1=X
360 NEXT
370 RETURN
400 REM EVALUATING POLYNOMIAL
```

```
410 Y=A(0):FOR I=1 TO N:Y=Y*X+A(I):NEXT
420 RETURN
500 REM FINE TUNING SOLUTION
510 PRINT "FINE TUNING SOLUTION";B$="  "
520 IF ABS(Y1)<D THEN X=X1:GOTO 590
530 IF ABS(Y2)<D THEN X=X2:GOTO 590
540 Z=X:X=(X1+X2)/2:GOSUB 400
550 IF ABS(Y)<D THEN 590
560 IF ABS(Z-X)<D THEN B$="PROBABLY ":GO
TO 590
570 IF Y*Y2>0 THEN X2=X:GOTO 540
580 X1=X:GOTO 540
590 PRINT:PRINT "THERE IS " B$ "A ROOT A
T ";X:TEST=0:X=B
600 RETURN

READY.
```

Occasionally the program will say that there is PROBABLY a root at some number. To check if it is a root, or to check how close it is to a root, type the following after typing N to another go.

X = Z:GOSUB 400:PRINT Y

The value of the number printed tells you how close you are to a root.

Newton's method

The method of finding roots of equations given in the previous section can be improved by using the so-called Newton's method.

Suppose we want to find a root of the polynomial equation

$9*X^5 - 3*X^4 + X^3 - X^2 + 5*X - 4 = 0$

Denote the polynomial function by P(X) and let P'(X) be the following polynomial.

$5*9*X^4 - 4*3*X^3 + 3*X^2 - 2*X + 5.$

In fact P'(X) is the *derivative* of P(X), but this need not concern us. Now, if Y is approximately a root of the equation P(X) = 0 then the following

Y − P(Y)/P'(Y)

is usually a better approximation provided that P'(Y) is non-zero.

The next program uses this technique to find roots of polynomial equations.

```
10 REM POLY ROOTS VIA NEWTON
20 PRINT CHR$(147) CHR$(154) "ROOTS OF P
OLYNOMIALS BY NEWTON'S METHOD. "
30 PRINT "THIS PROGRAM ATTEMPTS TO FIND
ROOTS OF"
40 PRINT "POLYNOMIALS SUCH AS" CHR$(17)
50 PRINT "A*X^N + B*X^(N-1) + ... + C*X
 + D = 0.   "
60 INPUT "DEGREE OF POLYNOMIAL ";N:PRINT
70 IF N<2 OR N<>INT(N) THEN PRINT "INTEG
ER PLEASE.":GOTO 60
80 DIM A(N):PRINT "ENTER COEFFICIENTS TE
RM BY TERM." CHR$(17)
90 FOR I=0 TO N
100 IF I<N THEN PRINT "   COEFFICIENT OF
 X^";MID$(STR$(N-I),2);
110 IF I=N THEN PRINT "CONSTANT COEFFICI
ENT";
120 INPUT A(I)
130 IF A(0)=0 THEN PRINT "NOT ZERO PLEAS
E.":GOTO 100
140 NEXT
150 PRINT:PRINT "ENTER A GUESS VALUE FOR
 A ROOT." CHR$(17)
160 INPUT "GUESS ";X:PRINT
170 REM CALCULATING P'(X)
180 DIM B(N):FOR I=0 TO N:B(I)=(N-I)*A(I
):NEXT
190 GOSUB 500 : REM FINDING ROOT
200 REM ENDING
210 PRINT:PRINT CHR$(154),"ANOTHER GO? Y
 OR N"
220 GET G$:IF G$<>"Y" AND G$<>"N" THEN 2
20
230 IF G$="Y" THEN RUN
240 PRINT CHR$(147) "BYE FOR NOW":END
400 REM EVALUATING POLYNOMIAL
410 Y=A(0):FOR I=1 TO N:Y=Y*X+A(I):NEXT
420 Y1=B(0):FOR I=1 TO N-1:Y1=Y1*X+B(I):
NEXT
430 RETURN
```

```
500 REM FINDING SOLUTION
510 J=1:D=1E-9:GOSUB 400
520 IF ABS(Y)<D THEN 590
530 IF Y1=0 THEN PRINT "DIVISION BY ZERO
 - SORRY - MAKE ANOTHER GUESS.":RETURN
540 Z=X:X=X-Y/Y1:J=J+1:GOSUB 400
550 IF ABS(Y)<D THEN 590
560 IF ABS(Z-X)<D THEN B$="PROBABLY ":GO
TO 590
570 IF J>1000 THEN PRINT "SORRY, CAN'T F
IND A ROOT.":RETURN
580 GOTO 530
590 PRINT:PRINT "THERE IS " B$ "A ROOT A
T ";X
600 RETURN
```

Exponential function

Functions like 2^X, 10^X are called *power functions*, because the variable X occurs as a power. Power functions are accessed on the Commodore 64 by using PRINT 2↑X etc. There is one important power function that the Commodore 64 singles out; the exponential function EXP(X).

The exponential function is based on powers of the number E which has the value 2.71828183 approximately. Do not confuse this E with the E that appears when numbers are printed using scientific notation. The number E itself is defined by the following:

$$E = 1 + \frac{1}{1!} + \frac{1}{2!} + \frac{1}{3!} + \frac{1}{4!} + \frac{1}{5!} + \frac{1}{6!} + \ldots$$

where ... means that the sum carries on forever, and the symbol ! stands for factorial which is defined by

$$N! = N * (N-1) * (N-2) * \ldots * 2 * 1$$

that is, the product of the integers from 1 to N. For instance, 4! is 4*3*2*1 which is 24.

The exponential function is defined by

$$EXP(X) = E^X$$
$$= E↑X.$$

So that, in particular, E = EXP(1). Try typing the following on your 64.

    E = EXP(1)
    PRINT E↑5, EXP(5)

Because of the properties of power functions we have the following properties of the exponential function.

    EXP(X)*EXP(Y) = EXP(X+Y)
    EXP(X)/EXP(Y) = EXP(X−Y)

There is a very straightforward formula which may be used to calculate EXP(X) for any number X. This is given as follows:

$$EXP(X) = 1 + \frac{X}{1!} + \frac{X^2}{2!} + \frac{X^3}{3!} + \frac{X^4}{4!} + \frac{X^5}{5!} + \frac{X^6}{6!} + \ldots$$

Logarithmic function

What number X satisfies EXP(X) = 3? Since EXP(1) = 2.71828183 we see that X is just over 1. In fact the answer is 1.09861229 approximately.

The number X that satisfies EXP(X) = N is called the (natural) logarithm of N. It is usually denoted by LN(N). But, in common with most microcomputers, the Commodore denotes it by LOG(N).

The logarithmic function has the following properties.

    LOG(X*Y) = LOG(X) + LOG(Y)
    LOG(X/Y) = LOG(X) − LOG(Y)
    LOG(X^N) = N*LOG(X)

It is because of these properties that logarithms are important when performing multiplication, division etc., without computers.

The following relations hold between the exponential and logarithmic function.

    EXP(LOG(N)) = N
    LOG(EXP(N)) = N

One simple use of the logarithmic function is testing of large numbers. For instance, the number X satisfies the relation

    X < 10↑N

if and only if the following relation is satisfied

    LOG(X) < N*LOG(10).

Chapter 4 Powers

Such a reformulation is useful because the number 10↑N itself may cause an OVERFLOW ERROR.

Roots of other functions

Given a function such as X*EXP(X)+1, a root of the function is a number which when substituted into the function gives 0.

Two programs for finding the roots of quadratic equations were given earlier on. The first of these can be adapted for finding roots of other functions. When the program is RUN you will be asked to type in two lines:

    100 DEF FNA (X) = enter function of X here
    GOTO 100

Enter the function whose root you want in the first line.

```
10 REM ROOTS OF FUNCTIONS
20 PRINT CHR$(147),CHR$(154) " ROOTS OF
FUNCTIONS" CHR$(17)
30 PRINT "THIS PROGRAM ATTEMPTS TO FIND
ROOTS OF  FUNCTIONS." CHR$(17)
40 PRINT "TYPE FUNCTION OF X IN THE FOLL
OWING FORM "
50 PRINT CHR$(158) "100 DEF FNA(X)=X*EXP
(X)" CHR$(17)
60 PRINT CHR$(154) "THEN TYPE" CHR$(17)
70 PRINT CHR$(158) "GOTO 100" CHR$(154):
END
100 DEF FNA(X)=X*EXP(X)*LOG(ABS(X)+1)
150 PRINT:PRINT "ENTER RANGE OVER WHICH
SEARCH OF ROOTS"
160 PRINT "IS TO BE ATTEMPTED." CHR$(17)
170 INPUT "LOWER VALUE ";A
180 INPUT "UPPER VALUE ";B:PRINT
190 IF A>=B THEN PRINT "FIRST VALUE SHOU
LD BE LOWER.":GOTO 170
200 REM SEARCH
210 S=B-A:T=0:TEST=-1:D=1E-9
220 PRINT "TEST RUN";T+1;"**";10^T;"DIVI
SIONS **":GOSUB 300
230 IF TEST THEN S=S/10:T=T+1:IF T<4 THE
N GOTO 220
240 IF T=4 THEN PRINT:PRINT "CANNOT LOCA
```

```
TE ROOTS - SORRY."
250 PRINT:PRINT CHR$(154),"ANOTHER GO? Y
 OR N"
260 GET G$:IF G$<>"Y" AND G$<>"N" THEN 2
60
270 IF G$="Y" THEN RUN
280 PRINT CHR$(147) "BYE FOR NOW":END
300 REM STEP BY STEP SEARCH
310 X=A:Y=FNA(X):X1=A:Y1=Y
320 FOR X=A+S TO B STEP S
330 Y=FNA(X):X2=X:Y2=Y
340 IF Y1*Y2<=D THEN GOSUB 500
350 Y1=Y2:X1=X
360 NEXT
370 RETURN
500 REM FINE TUNING SOLUTION
510 PRINT "FINE TUNING SOLUTION":B$=""
520 IF ABS(Y1)<D THEN X=X1:GOTO 590
530 IF ABS(Y2)<D THEN X=X2:GOTO 590
540 Z=X:X=(X1+X2)/2:Y=FNA(X)
550 IF ABS(Y)<D THEN 590
560 IF ABS(Z-X)<D THEN B$="PROBABLY ":GO
TO 590
570 IF Y*Y2>0 THEN X2=X:GOTO 540
580 X1=X:GOTO 540
590 PRINT:PRINT "THERE IS " B$ "A ROOT A
T ";X:TEST=0:X=B
600 RETURN

READY.
```

If you get the PROBABLY message, stop the program and type

   PRINT FNA(Z)

to find out how close you are to a root.

CHAPTER 5
Sequences

Sequences (and series) are important concepts that appear all over the place. A *sequence* is just a list of number, for instance:

1, 2, 3, 4, 5, 19, 7, 8, 12
20, 18, 16, 14, 12, 10, 8
1, 0.1, 0.01, 0.001, 0.0001, 0.00001, 0.000001

The individual numbers or members of the sequence are called the *terms* of the sequence. Usually the sequence is created with some rhyme or reason, such as the last two above. The second one was created by using the formula 22 − 2*N for N = 1 to 7, while the last used $10/10^N$ for N = 1 to 7. Your Commodore 64 is good at creating sequences. The following simple program illustrates this. Insert your own formula (involving N) in the second line.

```
10 REM SEQUENCE GENERATOR
20 DEF FNA(N)= (insert formula involving N here)
30 FOR N = 1 TO 10
40 PRINT FNA(N);: IF N< 10 THEN PRINT CHR$(157); ",";
50 NEXT
60 PRINT
```

Here are some sequences produced by this program for various different formulae. Can you see what formula was used in each case? Check your guess by inserting the formula in the above program. (The answers are given later on in this section and further examples are given in subsequent sections.)

(a) 1, 6, 11, 16, 21, 26, 31, 36, 41, 46
(b) 1, 2, 4, 8, 16, 32, 64, 128, 256, 512
(c) 1, 4, 9, 16, 25, 36, 49, 64, 81, 100
(d) 1, 2, 4, 7, 11, 16, 22, 29, 37, 46
(e) 4, 4, 8, 12, 20, 32, 52, 84, 136, 220

The formulae for the first few sequences are not too difficult to determine. For (a) it is 5*N−4, (for (b) it is 2↑N/2 while for (c) it is N*N. The fourth one (d) is not quite so easy to guess, it is (N*N − N

+2)/2. Finally, the formula for the fifth one (e) is impossible to guess unless you've seen it before, in fact it is:

$$4 * INT(((0.5 + SQR(5)/2) \uparrow N - (0.5 - SQR(5)/2) \uparrow N)/SQR(5)).$$

Here are some other formulae that you might like to try out.

$1 + (-1) \uparrow N$
$N * (-1) \uparrow N$
$INT(SIN(N) * 10)$

Arithmetic sequences

An *arithmetic* sequence or arithmetic progression is a sequence in which each term of the sequence is the sum of the preceding term and a constant. Sequence (a) from the previous section is an example of an arithmetic sequence. The general formula for an arithmetic sequence is given by

$$A + (N-1)*D$$

where A is the first term of the sequence and D is the common difference. Here are some further examples of arithmetic sequences.

5, 10, 15, 20, 25, 30, 35, 40, 45, 50	(A = 5, D = 5)
1, 1.5, 2, 2.5, 3, 3.5, 4, 4.5, 5, 5.5	(A = 1, D = 0.5)
0, 2, 4, 6, 8, 10, 12, 14, 16, 18	(A = 0, D = 2)

The following program may help you analyse arithmetic sequences. You enter the first term of the sequence, the common difference and the number of terms required. Notice that a formula is not required since the Commodore 64 does the calculation iteratively. In addition the program adds up all the terms in the sequence and gives you the answer.

```
10 REM ARITHMETIC SEQUENCES
20 PRINT CHR$(147),"ARITHMETIC SEQUENCES
" CHR$(17)
30 PRINT "THIS CREATES ARITHMETIC SEQUEN
CES." CHR$(17)
40 INPUT "FIRST TERM ";A:PRINT
50 INPUT "COMMON DIFFERENCE ";D:PRINT
60 INPUT "NUMBER OF TERMS ";N:PRINT
70 IF N<1 OR N<>INT(N) THEN 60
80 REM THE SEQUENCE
90 PRINT CHR$(5) "THE SEQUENCE:"
100 TERM=A:SUM=0
```

```
110 FOR I=1 TO N
120 IF 38-POS(0)<LEN(STR$(TERM)) THEN PR
INT
130 PRINT TERM;: IF I<N THEN PRINT CHR$(1
57) ",";
140 SUM=SUM+TERM: TERM=TERM+D
150 NEXT I
160 PRINT:PRINT:PRINT CHR$(159);"THE SUM
 IS ";SUM
170 PRINT:PRINT CHR$(154),"ANOTHER GO? Y
 OR N"
180 GET G$: IF G$<>"Y" AND G$<>"N" THEN 1
80
190 IF G$="Y" THEN RUN
200 PRINT CHR$(147) "BYE FOR NOW":END

READY.
```

Which would you prefer?

Suppose the publisher of this book offered you a job and then asked how you would like to be paid: "Which would you prefer? Start at £3000 per six months with a rise of £120 after every 6 months or start at £6120 a year with a rise of £240 after every year." The amount of money received each year with either choice follows an arithmetic progression. But one choice is far better than the other — the first choice. Can you see why? Look at the following calculations:

	First offer	Second offer
First year	(First 6 months £3000)	
	(Second 6 months £3120)	
	£6120	£6120
Second year	(First 6 months £3240)	
	(Second 6 months) £3360	
	£6600	£6360
Third year	(First 6 months £3480)	
	(Second 6 months £3600)	
	£7080	£6600

You can see that if you were to stay at the job for more than one year then the first offer is better. Notice that in the first offer the amount received during the first 6 months each year increases by £240, and so the annual increase in in fact £480. The annual salary for the first offer fits into the arithmetic sequence with formula

$$6120 + (N-1)*480,$$

while the second has the formula

6120 + (N−1)*240

Suppose you were given a third alternative: "Start with £1440 a quarter with a rise of £60 at the end of every 3 months." which offer would you prefer now? Hopefully the answer should be clear to you. The calculation would go as follows:

	Third offer	
First year	(First 3 months £1440)	
	(Second 3 months £1500)	
	(Third 3 months £1560)	
	(Fourth 3 months £1620)	
		£6120
Second year	(First 3 months £1680)	
	(Second 3 months) £1740	
	(Third 3 months) £1800	
	(Fourth 3 months) £1860	
		£7080

If you compare any 3 monthly period from year to year then the increase is £240, but you get this every 3 months. Thus, with this new offer the annual increase is £960.

Geometric sequences

Another common type of sequence is the *geometric* sequence or geometric progression. In a geometric sequence the ratio proceeds with a constant ratio, for instance

2, 6, 18, 54, 162

where every term (except the first) is three times the previous term. The general formula for a geometric sequence is given by

A * R↑(N−1)

where A is the first term and R is the common ratio.

Here are some further examples of geometric sequences.

4, 2, 1, 0.5, 0.25, 0.125, 0.0625   (A = 4, R = 0.5)
2, −4, 8, −16, 32, −64, 128   (A = 2, R = −2)

The next program may help you analyse geometric sequences. You enter the first term of the sequence, the common ratio and the number of terms required. Notice that a formula is not required since the Commodore 64 does the calculation iteratively. In addition the program adds up all the terms in the sequence and gives you the answer.

Chapter 5 Sequences

```
10 REM GEOMETRIC SEQUENCES
20 PRINT CHR$(147)," GEOMETRIC SEQUENCES
 " CHR$(17)
30 PRINT "THIS PROGRAM CREATES GEOMETRIC
 SEQUENCES "
40 INPUT "FIRST TERM ";A:PRINT
50 INPUT "COMMON RATIO ";R:PRINT
60 INPUT "NUMBER OF TERMS ";N:PRINT
70 IF N<1 OR N<>INT(N) THEN 60
80 REM THE SEQUENCE
90 PRINT CHR$(5) "THE SEQUENCE:" CHR$(17
)
100 TERM=A:SUM=0
110 FOR I=1 TO N
120 IF 38-POS(0)<LEN(STR$(TERM)) THEN PR
INT
130 PRINT TERM;:IF I<N THEN PRINT CHR$(1
57) ",";
140 SUM=SUM+TERM:TERM=TERM*R
150 NEXT I
160 PRINT:PRINT:PRINT CHR$(159);"THE SUM
 IS ";SUM
170 PRINT:PRINT CHR$(154),"ANOTHER GO? Y
 OR N"
180 GET G$:IF G$<>"Y" AND G$<>"N" THEN 1
80
190 IF G$="Y" THEN RUN
200 PRINT CHR$(147) "BYE FOR NOW":END

READY.
```

Interest

On 1 January a woman puts £100 in a bank which gives 6% interest each year (at the end of the year). To what amount will the woman's £100 grow after 10 years in the bank?

After one year she will have

$$100 + 100 * 6/100$$
$$= 100 + 100 * 0.06$$
$$= 100 * 1.06$$

which is £106. At the end of two years she will have

$$100 * 1.06 + (100 * 1.06) * 0.06$$
$$= 100 * 1.06 * 1.06$$

and you should be able to observe that after 10 years she will have

100 * 1.06↑10.

The total amount at the end of each year forms a geometric sequence as shown below.

100 * 1.06, 100 * 1.06↑2, 100 * 1.06↑3, 100 * 1.06↑4,
100 * 1.06↑5, 100 * 1.06↑6, 100 * 1.06↑7, 100 * 1.06↑8,
100 * 1.06↑9, 100 * 1.06↑10

More generally, if you start with an amount A and receive interest I% per annum then after N years your original amount has grown to the following amount.

A * (1 + I/100)↑N

Daily interest

If £1000 is deposited in a savings bank paying 6% interest at the end of each year then at the end of one year the total will become

1000 * 1.06

assuming that no further deposits or withdrawals are made. If, on the other hand, the bank paid interest every 6 months (and paid interest on the interest given, ie compounded the interest) then the total at the end of the year would be

1000 * (1.03)↑2.

More generally, if the bank paid 6% interest compounded N times a year then at the end of one year the £1000 would grow to the following amount.

1000 * (1 + 0.06/N)↑N

The table below illustrates the different amounts depending on how often interest is compounded.

N	6% compounded	Total at end of year (to nearest penny)
1	yearly	£1060.00
2	semiannually	£1060.90
4	quarterly	£1061.36
6	bimonthly	£1061.52
12	monthly	£1061.68

Chapter 5 Sequences

```
  52    weekly              £1061.80
 365    daily               £1061.83
8760    hourly              £1061.84
```

The above table was prepared with the Commodore 64 using the following simple program.

```
10 REM COMPOUND INTEREST VIA FORMULA
20 PRINT "ENTER NUMBER OF TIMES INTEREST IS TO BE COMPOUNDED"
30 INPUT "NUMBER";N : IF N<>INT(N) THEN 30
40 T = 1000 * (1 + 0.06/N)↑N : ? INT(T*100)/100
50 ?,"ANOTHER GO? Y OR N"
60 GET G$:IF G$<>"Y" AND G$<>"N" THEN 60
70 IF G$ = "Y" THEN RUN
```

You could make some additions and alterations to allow for other interest rates. Changes and additions in the next listing are marked by an asterisk at the beginning of a line.

```
 10 REM COMPOUND INTEREST VIA FORMULA - VARIABLE INTEREST
*15 INPUT "INTEREST RATE ";I: IF I<=0 OR I>=100 THEN 15
 20 PRINT "ENTER NUMBER OF TIMES INTEREST IS TO BE COMPOUNDED"
 30 INPUT "NUMBER";N : IF N<>INT(N) OR N<1 THEN 30
*35 IF N>20000 THEN PRINT "WARNING * ANSWER MAY BE INACCURATE"
*40 T = 1000 * (1 + I/100/N)↑N : ? INT(T*100)/100
 50 ?,"ANOTHER GO? Y OR N"
 60 GET G$:IF G$<>"Y" AND G$<>"N" THEN 60
 70 IF G$="Y" THEN RUN
```

Line 35 has been included because of inaccuracies that arise as a consequence of the large numbers involved.

Double or quit

Some people believe that you need never lose when gambling.

To illustrate this look at the following gambling game: "A fair coin is tossed, meanwhile you place your bet. If the coin shows a head then you get your money back plus an equivalent amount."

To show that you need never lose the argument goes as follows. Start by betting £1. If you win, quit. If you lose, play again with a stake of £2. Each time you lose, double your stake and play again. Stop as soon as you win and you will be in pocket.

Suppose for example that you lose the first four times and win on the fifth. The table below illustrates what happens.

	Stake	Loss	Win
First toss	£1	£1	
Second toss	£2	£2	
Third toss	£4	£4	
Fourth toss	£8	£8	
Fifth toss	£16		£16
TOTAL		£15	£16

NET GAIN = £1

The sequence that arises is a geometric sequence. Do you believe the argument over why you would never lose?

Fibonacci sequences

At the beginning of this chapter we had the following sequence.

(e) 4, 4, 8, 12, 20, 32, 52, 84, 136, 220

which is given by the following formula

4*INT (((0.5 + SQR(5)/2)↑N − (0.5 − SQR(5)/2)↑N) / SQR(5)).

Rather than use this formula there is a more obvious way of creating the sequence. Every term except the first two is the sum of the two previous ones.

$4 + 4 = 8$
$4 + 8 = 12$
$8 + 12 = 20,$

and so on.

Sequences of numbers created in this way are called Fibonacci sequences. It was in 1202 that Leonard of Pisa, nicknamed Fibonacci, observed such a sequence of numbers associated with the breeding of rabbits.

Here are two other Fibonacci sequences:

2, 5, 7, 12, 19, 31, 50
3, 5, 8, 13, 21, 34, 55

The next program will produce Fibonacci sequences ad nauseum.
```
10 REM FIBONACCI SEQUENCES
20 PRINT CHR$(147)," FIBONACCI SEQUENCES
" CHR$(17)
```

Chapter 5 Sequences

```
30 PRINT "THIS PROGRAM CREATES FIBONACCI
 SEQUENCES "
40 PRINT "TYPE IN TWO INTEGERS, SEPARATE
D BY A     COMMA." CHR$(17)
50 INPUT "NUMBERS ";U,V:PRINT
60 IF U<>INT(U) OR V<>INT(V) THEN PRINT
"INTEGERS PLEASE":GOTO 50
70 INPUT "HOW MANY TERMS DO YOU WANT ";N
:PRINT
80 IF N<1 OR N<>INT(N) THEN 70
90 REM THE SEQUENCE
100 PRINT CHR$(5) "THE FIBONACCI SEQUENC
E:" CHR$(17)
110 FOR I=1 TO N
120 IF 38-POS(0)<LEN(STR$(U)) THEN PRINT
130 PRINT U;:IF I<N THEN PRINT CHR$(157)
 ",";
140 W=U+V:U=V:V=W
150 NEXT I
160 PRINT:PRINT:PRINT CHR$(154),"ANOTHER
 GO? Y OR N"
170 GET G$:IF G$<>"Y" AND G$<>"N" THEN 1
70
180 IF G$="Y" THEN RUN
190 PRINT CHR$(147) "BYE FOR NOW":END

READY.
```

Here is an exercise that you may like to do. Use your Commodore 64 and write a short program.

"Write down any two integers. Form the Fibonacci sequence by adding pairs of terms to form a third term. Find the ratio of each term in the sequence with the one immediately before it. What happens to this ratio as the number of terms gets large? Work out the value of 0.5 + SQR(5)/2."

CHAPTER 6
Number Bases

We usually record numbers using the *decimal system* of notation. For instance 1432, which we call one thousand four hundred and thirty two, stands for the more awkward expression

$1*10^3 + 4*10^2 + 3*10 + 2.$

We can rewrite this in a slightly more awkward way:

$1*10^3 + 4*10^2 + 3*10^1 + 2*10^0$

since $10^1 = 10$, and $10^0 = 1$. In other words, the number 1432 is interpreted as a sum of multiples of powers of 10. The integers 1, 4, 3, and 2 are called the *digits* of the number with 1 being the thousands digit, 4 the hundreds digit, 3 the tens digit and 2 the units digit. Technically we refer to this representation of the number as its *decimal representation* and say that the number is expressed to the *base* of 10. The word decimal comes from the Latin *decem*, ten.

The decimal system has a base of 10. But bases other than 10 can be used. Using different bases to interpret numbers is both interesting and useful. For example, numbers represented in base 2 have proved to be extremely important in computers and computer related activities.

Any integer greater than 1 can be used as a base, and any number can be expressed in any base. Furthermore, it is easy for your Commodore 64 to convert a number expressed in one base into another base.

Let N stand for any positive integer, and let B be an integer greater than 1. To express the number N to the base B we need to write N in the following way:

$N = X_m*B^m + X_{m-1}*B^{m-1} + \ldots + X_1*B + X_0$

where each of the numbers $X_0, X_1, \ldots, X_m$ are integers between 0 and $B-1$. (See what happens if you substitute 10 for B.) The digits X_0, X_1, etc, are called the *coefficients* of the number N to base B.

Small values of B, the base, give long representations of the numbers. But they have the advantage of requiring fewer choices for the coefficients. The extreme case occurs when $B = 2$. The resulting system is called the *binary* number system (from the Latin *binarius*, two). When a number is written in the binary system only the integers 0 and 1 can appear as coefficients. For example

$$86 = 64 + 16 + 4 + 2$$
$$= 1*2^6 + 0*2^5 + 1*2^4 + 0*2^3 + 1*2^2 + 1*2 + 0$$

Thus the number 86 expressed in binary form is 1010110. Binary numbers are used by computers because they are represented as strings of zeros and ones. The reason is that 0 and 1 can be easily expressed in a computer by a switch being either off or on.

For bases larger than 10 we need some extra symbols. The obvious symbols to use are the letters of the alphabet A, B, C, etc. A common base that is used is 16. A number expressed in the base 16 is called *hexa-decimal*. The advantage of this base is that it requires few coefficients to express a number and yet hexa-decimal numbers are easily converted to binary numbers.

To convert a number from base 10 to base B is quite straightforward. Suppose we want to convert the number N from base 10 to base B. First subtract all multiples of B from N.

$$M = INT(N/B) : R = N - B * M$$

Record the remainder and call it R_0. Now repeat the process with M by setting N = M. Call the new remainder R_1. Continue in this way until the value of M reaches 0. Suppose that R_S is the last remainder we find, then the original number N to base B is

$$R_S \ldots R_2 R_1 R_0$$

Let's go through a specific example. Suppose we want to convert the number 29 to base 3. The calculation proceeds as follows.

Step 1.
$$N = 29$$
$$M = INT(29/3)$$
$$= 9$$
$$R_0 = 29 - 3*9$$
$$= 2$$

Step 2.
$$N = M$$
$$= 9$$
$$M = INT(9/3)$$
$$= 3$$
$$R_1 = 9 - 3*3$$
$$= 0$$

Chapter 6 Number Bases

Step 3.
    N = M
      = 3
    M = INT(3/3)
      = 1
    R_2 = 3 − 3*1
      = 0

Step 4.
    N = M
      = 1
    M = INT(1/3)
      = 0
    R_3 = 1 − 3*0
      = 1

The process stops after four steps when M reaches 0. The value of 29 to base 3 is 1002.

The next program converts integers from one base to another base. For example you could convert numbers in base 10 to base 16. For bases greater than 10 the letters A, B, C, etc are used to represent the numbers 10, 11, 12, etc.

```
10 REM BASE CONVERTER
20 PRINT CHR$(147)," BASE CONVERTER" CH
R$(17)
30 PRINT "THIS PROGRAM CONVERTS INTEGERS
 FROM ONE BASE TO ANOTHER." CHR$(17)
40 INPUT "ENTER BASE TO CONVERT FROM";A:
PRINT CHR$(17);
45 AA=54+A:IF A<11 THEN AA=47+A
50 IF A<2 OR A>35 OR INT(A)<>A THEN PRIN
T "SILLY - TRY AGAIN":GOTO 40
60 INPUT "ENTER NUMBER TO BE CONVERTED";
N$:PRINT CHR$(17);
70 REM CHECK N$ IS OF THE RIGHT FORM
80 IF N$="" THEN PRINT,"NOT A NUMBER":GO
TO 60
90 L=LEN(N$):I=0
100 I=I+1:N=ASC(MID$(N$,I,1))
110 IF N<48 OR (N>57 AND N<65) OR N>AA T
HEN PRINT,"NOT A NUMBER":GOTO 60
120 IF I<L THEN 100
122 REM STORE N$ IN ARRAY
124 DIM A(L)
125 FOR I=1 TO L:N=ASC(MID$(N$,I,1)):IF
```

Mathematics on the Commodore 64

```
N<58 THEN A(I)=N-48
126 IF N>64 THEN A(I)=N-55
127 NEXT
150 REM CONVERT NUMBER FROM BASE A TO BA
SE 10
160 N=VAL(N$)
170 IF A<>10 THEN N=0:FOR I=1 TO L:N=A(I
)+N*A:NEXT
180 PRINT "DECIMAL FORM OF NUMBER" N CHR
$(17)
190 INPUT "ENTER BASE TO CONVERT TO":B:P
RINT CHR$(17);
200 IF B<2 OR B>35 OR INT(B)<>B THEN PRI
NT "SILLY - TRY AGAIN":GOTO 190
210 REM CONVERT N TO BASE B
220 N$=""
230 M=INT(N/B):R=N-B*M:N=M
240 IF R<10 THEN N$=CHR$(48+R)+N$
250 IF R>9 THEN N$=CHR$(55+R)+N$
260 IF N<>0 THEN 230
270 PRINT "NUMBER TO BASE" B "IS " N$ CH
R$(17)
280 PRINT,"ANOTHER GO? Y OR N"
290 GET G$:IF G$<>"Y" AND G$<>"N" THEN 2
90
300 IF G$="Y" THEN RUN
310 PRINT CHR$(147) "BYE FOR NOW":END

READY.
```

64 numbers

A number between 0 and 255 can be represented as a binary number using at most 8 coefficients. For example

$$255 = 1*2^7 + 1*2^6 + 1*2^5 + 1*2^4 + 1*2^3 + 1*2^2 + 1*2 + 1$$
$$128 = 1*2^7 + 0*2^6 + 0*2^5 + 0*2^4 + 0*2^3 + 1*2^2 + 0*2 + 0$$

These 8 coefficients, or 8 bits on a computer is called a byte. The Commodore 64 stores integers using two bytes called the high and low bytes. The high byte represents multiples of 256. For example the number 999 would be stored with high byte 3 and low byte 231 since 999 = 3*256 + 231. Storing numbers using bytes is the same as storing numbers to the base of 256.

Chapter 6 Number Bases

You can see how the 64 stores the high and low byte of an integer by PEEKing. Recall that Integer variables on the Commodore 64 are specified by the percent (%) sign after a variable name. Type the following, pressing return at the end of each line.

NEW : CLR
X% = 999   (type any integer here)
PRINT "HIGH BYTE=" PEEK(2053), "LOW BYTE=" PEEK(2054)

The largest integer that the Commodore 64 can store is 32767 which equals 127*256 + 255. Numbers with a high byte of 128 or larger are negative numbers. Indeed negative numbers are stored by first looking at the ABSolute value of the number, calculating the high and low order bytes and then subtracting the value of the high byte from 255 and the low byte from 256. Thus, for example, −1 would have a high byte of 255 and a low byte of 255.

The following shows you how to calculate a number from the high and low order byte. Let H be the high order byte and L the low order byte.

NUMBER = H*256 + L
IF H> = 128 THEN NUMBER = −((255−H)*256 + (256−L))

The last line is equivalent to the following line.

IF H > = 128 THEN NUMBER = H*256 + L − 256*256

Try POKEing numbers into locations 2053 and 2054, then get the 64 to PRINT X% and compare the answers. For example try typing the following, pressing return at the end of each line.

NEW : CLR : X% = 0
POKE 2053,98 : POKE 2054,99   (replace 98 and 99)
N = PEEK(2053)*256 + PEEK(2054)
IF PEEK(2053) > = 128 THEN N = N − 256*256
PRINT X%,N

Small numbers

The program in the section before the last one works on positive whole numbers. Any number, integral or non-integral, has a representation in any base. For example the decimal number 0.25 expressed in binary takes the form 0.01, while the decimal number 0.125 is 0.001 in binary. To see this first let's see what we mean by the decimal number 0.25. This number represents two-tenths and five-hundredths, that is

$0.25 = 2/10 + 5/100$
$ = 2*10^{-1} + 5*10^{-2}.$

69

Mathematics on the Commodore 64

To express it to the base B means writing it in the form

$$Y_1 * B^{-1} + Y_2 * B^{-2} + \ldots$$

where as usual $B^{-1} = 1/B$, $B^{-2} = 1/B^2$, etc.

The decimal numbers 0.25 and 0.125 may be written in the following way:

$$0.25 = 1/4$$
$$= 0*2^{-1} + 1*2^{-2}$$
$$0.125 = 1/8$$
$$= 0*2^{-1} + 0*2^{-2} + 1*2^{-3}$$

which explains the binary form of these numbers.

As another example look at the number 0.6 expressed in terms of negative) powers of 2.

$$0.6 = 1*2^{-1} + 0*2^{-2} + 0*2^{-3} + 1*2^{-4}$$
$$+ 1*2^{-5} + 0*2^{-6} + 0*2^{-7} + 1*2^{-8}$$
$$+ 1*2^{-9} + 0*2^{-10} + 0*2^{-11} + 1*2^{-12}$$
$$+ 1*2^{-13} + 0*2^{-14} + 0*2^{-15} + 1*2^{-16}$$
$$+ \ldots$$

In fact we need infinitely many terms to express 0.6 accurately in binary form which takes the form

$$0.1001100110011001100110011001\ldots$$

The Commodore 64 only stores 32 of these digits starting with the first non-zero one, in addition it rounds up if the thirty-third significant digit is non-zero. Thus the 64 stores 0.6 as

$$0.10011001100110011001100110011010$$

in binary form.

The next program displays the binary form, as stored by the Commodore 64, of a decimal number between 0 and 1.

```
10 REM DECIMAL TO BINARY
20 PRINT CHR$(147),"DECIMAL TO BINARY" C
HR$(17)
30 PRINT "THIS PROGRAM PRINTS THE BINARY
   FORM OF ANUMBER ";
40 PRINT "BETWEEN 0 AND 1." CHR$(17)
50 N=0:INPUT "TYPE IN YOUR NUMBER ";N:PR
INT
```

Chapter 6   Number Bases

```
60 IF N<=0 OR N>=1 THEN PRINT "BETWEEN 0
   AND 1 !!":GOTO 50
70 N$="0."
80 FOR I=1 TO 32
90 N=N*2:N$=N$+MID$(STR$(INT(N)),2,1):N=
N-INT(N)
100 NEXT I
110 PRINT "THE BINARY FORM OF YOUR NUMBE
R IS:" CHR$(17)
120 PRINT N$ CHR$(17)
130 PRINT,"ANOTHER GO? Y OR N"
140 GET G$:IF G$<>"Y" AND G$<>"N" THEN 1
40
150 IF G$="Y" THEN PRINT:GOTO 50
160 PRINT CHR$(147) "BYE FOR NOW":END

READY.
```

Floating points

Integers in the Commodore 64 are stored using 2 bytes. But numbers are stored using 5 bytes, even if the number itself is an integer. Unless you declare your number to be an integer by using the percent sign it will be stored as a real number using 5 bytes. A number can be expressed in binary form in the following way:

$$1.X_1X_2X_3\ldots X_m * 2^N$$

where $X_1, X_2, \ldots, X_m$ are either 0 or 1, and N is an integer (positive, negative or zero). The integer N is called the *binary exponent* of the number, the other part is called the *binary mantissa*. For instance decimal 10 is binary 1010 which may be rewritten as

$$1.01 * 2^3$$

so that 10 has binary exponent 3 and binary mantissa 1.01. As another example look at decimal 0.375 which is binary 0.011 and so may be written as

$$1.1 * 2^{-2}$$

and so decimal 0.375 has binary exponent -2 and binary mantissa 1.1.

We have said that the 64 uses 5 bytes to store its numbers. The first byte is the binary exponent plus 129. The remaining four bytes give the binary mantissa and the sign of the number. Since the first term in the binary mantissa is always 1 we do not need to store it —

we simply store all the digits to the right of the decimal place in the binary mantissa. The first bit of byte 2 stores the sign of the number, the remaining 31 bits in the last four bytes store the binary mantissa (ignoring the leading 1).

For example, decimal 10 would be stored as follows: The first byte is 129 plus the binary exponent 3, which totals 132. The first bit of the second byte would be 0 since the number is positive. The remaining 31 bits would be

0100000000000000000000000000000

since the binary mantissa of 10 is 1.01 and we ignore the leading 1. Thus the 32 bits for the last four bytes would be

00100000000000000000000000000000

which, when broken into four groups of 8, give

00100000 00000000 00000000 00000000

which in turn are 32, 0, 0, 0. Thus the 5 bytes used to store the decimal 10 would be 132, 32, 0, 0, 0.

We can reverse the process and find the number that the 64 is holding in 5 bytes. Suppose that a number N is stored with the five bytes P, Q, R, S, T. The following program lines calculate N from P, Q, R, S and T.

$X = 1$: IF $Q \geq 128$ THEN $Q = Q - 128$: $X = -1$
$N = X * 2^{P-129} * (1 + Q*2^{-7} + R*2^{-15} + S*2^{-23} + T*2^{-31})$

To actually see the 64 in action type the following lines, pressing return at the end of each line.

NEW : CLR
X = 10   (type whatever number you like here)
FOR I=0 TO 4 : PRINT "BYTE" I+1 "=" PEEK(2053+I) : NEXT I

In fact if you do this then you'll find that the Commodore 64 has some minor errors in the way it performs multiplication. The number $1 + 2^{-24}$ is stored (correctly) with the following 5 bytes

129, 0, 0, 0, 128

However, the same number written as $1*(1 + 2^{-24})$ is stored with the following 5 bytes

129, 0, 0, 0, 64

Chapter 6 Number Bases

In other words, according to the Commodore 64

$1 * (1 + 2\uparrow-24) = 1 + 2\uparrow-25$

Another way of displaying the same problem is as follows:

X = 1 + 2↑−24
PRINT X − X, X − 1*X

Alternatively, try the following

X = 1 + 2↑−24
X1 = 1*X
X2 = 1*X1
X3 = 1*X2
X4 = 1*X3
PRINT X,X1,X2,X3,X4

Problems of this nature appear not to occur if the middle 3 bytes of the 5 bytes used to store the number are not all 0. See what examples you can find.

73

CHAPTER 7
Days and Weeks

Days

Zeller's congruence is a complicated looking formula that calculates the day of the week (Sunday, Monday, etc) for any given date. Using this formula you could, for example, find out on which day of the week a person was born. And, if your memory is bad, you could find out on what day of the week a certain anniversary occurred.

The following is Zeller's formula:

$$A = INT(2.6*M - 0.1) + D + Y + INT(Y/4) + INT(C/4) - 2*C$$
$$X = A - 7*INT(A/7)$$

The number X is a number between 0 and 6, because all multiples of 7 smaller than A have been subtracted from A. These numbers represent the 7 days of the week as follows:

   0 : Sunday,
   1 : Monday,
   2 : Tuesday,
   3 : Wednesday,
   4 : Thursday,
   5 : Friday,
   6 : Saturday.

The numbers D, M, Y and C are defined as follows:

   D : the day of the month.
   M : the number of the month — but not the standard number. January and February are numbers 11 and 12 of the preceding year (affecting Y and possibly C described below). March is number 1, April is 2, May is 3, ..., and December is number 10.
   Y : the year in the century.
   C : the number of hundreds in the year, in other words, the first two digits in the year number.

For instance, if the date is 26th August 1983 then the standard way of expressing this is 26/08/1983. For Zeller's formula we use D = 26, M = 6, Y = 83 and C = 19.

Substituting these values into Zeller's formula gives:

$$A = INT(2.6*6 - 0.1) + 26 + 83 + INT(83/4) + INT(19/4) - 2*19$$
$$= INT(15.6 - 0.1) + 26 + 83 + INT(20.75) + INT(4.75) - 38$$
$$= 15 + 26 + 83 + 20 + 4 - 38$$
$$= 110$$

and

$$Y = 110 - 7*INT(110/7)$$
$$= 110 - 7*INT(15.714285)$$
$$= 110 - 105$$
$$= 5.$$

Thus we conclude that the day of the week of 26th August 1983 is Friday.

Here are some examples showing the standard date format and the values that Zeller's formula uses.

STANDARD NOTATION	D	M	C	Y
		ZELLER'S FORMULA NOTATION		
03/03/1947	3	1	19	47
01/01/2000	1	11	19	99
26/02/1983	26	12	19	82
29/11/1984	29	9	19	84

The next program uses Zeller's formula to calculate the day of the week for any specified date. Your Commodore 64 will automatically calculate the correct values of D, M, C and Y required from any date you input. Observe that the program makes a few checks to ensure that the date entered makes sense. Thus, for instance, 30th February 1983 will not be accepted. In addition the year entered must be an integer in the range 1752 to 4902. Zeller's formula applies in the range 1582 to 4902, but the Gregorian calendar has been used in Britain, the British Colonies and the USA only since 1752.

Leap years are automatically taken care of in the program. Note that a year is a leap year if the year number is exactly divisible by 4, unless it is divisible by 100 but not divisible by 400. Thus 1900 was not a leap year but 2000 will be a leap year.

```
10 REM DAY OF WEEK
20 DIM A(12),A$(12):FOR I=1 TO 12:READ A
(I),A$(I):NEXT I
30 FOR I=0 TO 6:READ B$(I):NEXT I
40 PRINT CHR$(147),"        DAY OF WEEK" CHR
$(17)
```

```
50 PRINT "THIS PROGRAM CALCULATES THE DA
Y OF THE"
60 PRINT "WEEK FOR ANY DATE SPECIFIED."
CHR$(17)
70 PRINT "TYPE IN THE DATE:" CHR$(17)
80 A(2)=29:D=0:INPUT "DAY, 1 TO 31 : ";D
:PRINT
90 IF D<1 OR D>31 OR D<>INT(D) THEN PRIN
T,"TRY AGAIN":GOTO 80
100 M=0:INPUT "MONTH, 1 TO 12 : ";M:PRIN
T
110 IF M<1 OR M>12 OR M<>INT(M) THEN PRI
NT,"TRY AGAIN":GOTO 100
120 IF D>A(M) THEN PRINT "NOT ENOUGH DAY
S IN MONTH, TRY AGAIN":GOTO 80
130 Y=0:INPUT "YEAR, IN FULL : ";Y:PRINT
140 IF Y<1582 OR Y>4902 THEN PRINT,"NOT
IN RANGE":GOTO 130
150 IF Y<>INT(Y) THEN PRINT,"NOT A YEAR,
 TRY AGAIN":GOTO 130
160 REM CHECK FOR LEAP YEAR
170 L=0:IF INT(Y/4)*4=Y THEN L=-1
180 IF L AND INT(Y/100)*100=Y THEN L=0:I
F INT(Y/400)*400=Y THEN L=-1
190 A(2)=28-L:IF L THEN PRINT "THIS IS A
 LEAP YEAR" CHR$(17)
200 IF M=2 AND D>A(2) THEN PRINT "28 DAY
S IN FEBRUARY":GOTO 80
210 PRINT D;A$(M);Y;"IS/WAS A ";
220 REM CHANGE FORMAT OF INPUT DETAILS
230 M=M-2:IF M<1 THEN M=M+12:Y=Y-1
240 C=INT(Y/100):Y=Y-C*100
250 REM ZELLER'S CONGRUENCE
260 DAY=INT(2.6*M-0.1)+D+Y+INT(C/4)+INT(
Y/4)-2*C
270 DAY=DAY-7*INT(DAY/7):PRINT B$(DAY)+"
DAY" CHR$(17)
280 PRINT,"ANOTHER GO? Y OR N"
290 GET G$:IF G$<>"Y" AND G$<>"N" THEN 2
90
300 IF G$="Y" THEN PRINT:GOTO 70
310 PRINT CHR$(147) "BYE FOR NOW":END
400 REM DATA
```

Mathematics on the Commodore 64

```
410 DATA 31,JANUARY,29,FEBRUARY,31,MARCH
,30,APRIL
420 DATA 31,MAY,30,JUNE,31,JULY,31,AUGUS
T
430 DATA 30,SEPTEMBER,31,OCTOBER,30,NOVE
MBER,31,DECEMBER
440 DATA SUN,MON,TUES,WEDNES,THURS,FRI,S
ATUR

READY.
```

Note: You may have come across Zeller's formula before and possibly noticed that the formula used here is slightly different. Often the first term INT(2.6*M − 0.1) is written as INT(2.6*M − 0.2) instead. This latter form is not used here because of the way the Commodore calculates the INTegral part of numbers. Try M = 7, then

$$\text{INT}(2.6*7 - 0.2) = \text{INT}(18.2 - 0.2)$$
$$= \text{INT}(18)$$
$$= 18$$

However, the Commodore 64 returns the value of INT(2.6*7 − 0.2) as 17, even though if you ask it to print 2.6*7 − 0.2 it prints 18 correctly.

Calendar

Once we know the day of the week of any date we can produce a calendar. The next program prints a calendar for any month in any year. (Only one month can be displayed reasonably on the screen.) The program calculates the day on which the first day of that month occurs by using Zeller's formula. The remaining days are then printed out. As in the Day of the Week program, leap years are automatically taken care of.

```
10 REM CALENDAR
20 DIM A(12),A$(12):FOR I=1 TO 12:READ A
(I),A$(I):NEXT I
30 REM THE START
40 PRINT CHR$(147),"          CALENDAR" CHR$
(17)
50 PRINT "THIS PROGRAM PRINTS A CALENDAR
 FOR ANY"
60 PRINT "MONTH OF ANY YEAR SPECIFIED."
CHR$(17)
70 PRINT "TYPE IN THE MONTH AND YEAR : "
CHR$(17)
```

```
80 M=0:INPUT "MONTH, 1 TO 12 : ";M:PRINT
90 IF M<1 OR M>12 OR M<>INT(M) THEN PRIN
T,"TRY AGAIN":GOTO 80
100 Y=0:INPUT "YEAR, IN FULL : ";Y:PRINT
110 IF Y<1752 OR Y>4902 THEN PRINT,"NOT
IN RANGE":GOTO 100
120 IF Y<>INT(Y) THEN PRINT,"NOT A YEAR.
 TRY AGAIN":GOTO 100
130 REM CHECK FOR LEAP YEAR
140 L=0:IF INT(Y/4)*4=Y THEN L=-1
150 IF L AND INT(Y/100)*100=Y THEN L=0:I
F INT(Y/400)*400=Y THEN L=-1
160 A(2)=28-L:IF L THEN PRINT "THIS IS A
 LEAP YEAR" CHR$(17)
170 A$=A$(M)+STR$(Y):MM=A(M)
200 REM CHANGE FORMAT OF INPUT DETAILS
210 M=M-2:IF M<1 THEN M=M+12:Y=Y-1
220 C=INT(Y/100):Y=Y-C*100
230 REM ZELLER'S CONGRUENCE
240 DAY=INT(2.6*M-0.1)+1+Y+INT(C/4)+INT(
Y/4)-2*C
250 DAY=DAY-7*INT(DAY/7)
260 REM PRINT CALENDAR
270 PRINT CHR$(147):PRINT:POKE 53281,7
280 PRINT TAB(20-LEN(A$)/2) CHR$(31) A$
CHR$(17)
290 PRINT CHR$(28) "   SUN   MON   TUE   WE
D   THU   FRI   SAT   " CHR$(31)
300 FOR I=1 TO MM
310 DAY=DAY+1
320 PRINT TAB(DAY*5-2+(I>9)) I;
330 IF DAY>6 THEN DAY=0:PRINT:PRINT
340 NEXT I
350 PRINT:PRINT:PRINT
360 PRINT,CHR$(5) " ANOTHER GO? Y OR N"
CHR$(31)
370 GET G$:IF G$<>"Y" AND G$<>"N" THEN 3
70
380 IF G$="Y" THEN PRINT:GOTO 70
390 PRINT CHR$(147) "BYE FOR NOW":END
400 REM DATA
410 DATA 31,JANUARY,29,FEBRUARY,31,MARCH
,30,APRIL
```

```
420 DATA 31,MAY,30,JUNE,31,JULY,31,AUGUS
T
430 DATA 30,SEPTEMBER,31,OCTOBER,30,NOVE
MBER,31,DECEMBER

READY.
```

Date management

Occasionally there is a need to provide a listing of dates that are a specified number of days apart. For instance, treatment days at a hospital, and pay days.

To produce such a listing we use the 'pseudo-Julian' date. This date is simply the number of days since some fixed date. (In fact the First of January of the year 1 has a pseudo-Julian date of 1). A relatively simple formula converts the real date to the pseudo-Julian date and vice-versa.

If the date is D/M/Y where D is the day, M the month number and Y the year (including the century) then the pseudo-Julian date is calculated as follows:

$X = INT(30.57*M) + INT(365.25*Y - 395.25) + D$
IF $M > 2$ and Y is a leap year then subtract 1 from X.
If $M > 2$ and Y is not a leap year then subtract 2 from X.

For example, let's calculate the pseudo-Julian date of 26th August 1983. The values of D, M and Y are given by $D = 26$, $M = 8$, $Y = 1983$. Substituting these values into the formula gives the following:

$X = INT(30.57*8) + INT(365.25*1983 - 395.25) + 26$
$= INT(244.56) + INT(723895.5) + 26$
$= 244 + 723895 + 26$
$= 724165$

However, since the month number M is greater than 2 and 1983 is not a leap year we subtract 2 from X to give a pseudo-Julian date of 724163.

To calculate the date from the pseudo-Julian date proceed as follows, where X is the pseudo-Julian date.

The first approximation to the year is given by:
$Y = INT(X/365.26) + 1$

the day within the year is given by:
$D = X - INT(365.25*Y - 395.25)$

A leap year adjustment is made:
$D1 = 2$, if it is a leap year then $D1 = 1$
If $D > 91 - D1$ then add D1 to D

Chapter 7 Days and Weeks

Calculate the month and day:
M = INT(D/30.57)
D = D − INT(30.57*M)

Adjust month and year if necessary:
If M > 12 then set M to 1 and add 1 to Y.

For example, let's calculate the date corresponding to a pseudo-Julian date of 724164 (this is 1 higher than the pseudo-Julian date that we calculated earlier on). The calculations are as follows:

Y = INT(724164/365.26) + 1
 = INT(1982.5987) + 1
 = 1983
D = 724164 − INT(365.25*1983 − 395.25)
 = 724164 − INT(724290.75 − 395.25)
 = 724164 − INT(723895.5)
 = 269

The year 1983 is not a leap year so that D1 is 2. The value of D is greater than 91 − D1 and so we add D1 to D. Thus the value of D is now given by

D = 271
M = INT(271/30.57)
 = INT(8.86490023)
 = 8
D = 271 − INT(30.57*8)
 = 271 − INT(244.56)
 = 271 − 244
 = 27

The value of M is not greater than 12 and so we are finished, with values of D = 27, M = 8 and Y = 1983. Thus the date corresponding to the pseudo-Julian date of 724164 is 27th August 1983.

The next program performs all the above sort of calculations quickly and provides a listing of dates that are a specified number of days apart. For convenience the program only works for dates in the 20th century. You should be able to make any changes necessary for another century quite easily.

Warning. Dates are entered in the form DD/MM/YY, for example 12th March 1984 would be entered as 12/03/84 or 12/ 3/84 but not as 12/3/84. Not too many checks have been made for the date entered, and you could, for instance, enter 30/02/83. The program would think of this date as 2nd March 1983 (can you see why?).

```
10 REM DATE MANAGEMENT
20 POKE 53281,7:PRINT CHR$(147),CHR$(28)
   "  DATE MANAGEMENT" CHR$(17)
30 PRINT CHR$(31) "THIS PROGRAM PROVIDES
   THE LISTING OF"
40 PRINT "DATES THAT ARE A SPECIFIED NUM
BER OF"
50 PRINT "DAYS APART." CHR$(17)
60 REM ENTER START DATE
70 PRINT "TYPE IN DATE AS DD/MM/YY, E.G.
   15/02/84 "
80 S$="":INPUT "ENTER START DATE: ";S$:P
RINT
90 IF LEN(S$)<>8 THEN PRINT,"IN THE FORM
   DD/MM/YY":GOTO 80
100 IF MID$(S$,3,1)<>"/" OR MID$(S$,6,1)
<>"/" THEN PRINT,"USE DD/MM/YY":GOTO 80
110 D=VAL(MID$(S$,1,2)):M=VAL(MID$(S$,4,
2)):Y=VAL(MID$(S$,7,2))+1900
120 IF D<=0 OR D>31 OR M<=0 OR M>12 THEN
   PRINT,"DAY, MONTH ERROR":GOTO 80
130 REM DATE FORM IS NOW IN REASONABLY C
ORRECT FORM
140 REM CALCULATE THE PSEUDO-JULIAN DAY
150 X=INT(30.57*M)+INT(365.25*Y-395.25)+
D
160 REM ADJUST FOR LEAP YEAR
170 IF M>2 THEN X=X-2:IF INT(Y/4)*4=Y TH
EN X=X+1
180 REM ENTER INTERVAL OF DAYS
190 PRINT "ENTER INTERVAL IN DAYS BETWEE
N DATES."
200 P=0:INPUT "INTERVAL : ";P:PRINT
210 IF P<=0 OR INT(P)<>P THEN PRINT,"TYP
E A NUMBER":GOTO 200
220 PRINT "NUMBER OF TIMES INTERVAL REQU
IRED."
230 N=0:INPUT "NUMBER : ";N:PRINT
240 IF INT(N)<>N THEN PRINT,"A WHOLE NUM
BER":GOTO 230
250 IF N<=0 OR N>100 THEN PRINT,"BE REAS
ONABLE":GOTO 230
260 PRINT CHR$(28) : REM COLOUR FOR PRIN
```

TING DATES
```
270 FOR I=0 TO N
280 :  Y=INT(X/365.26)+1 : REM YEAR
290 :  D=X-INT(365.25*Y-395.25)
300 :  REM LEAP YEAR ADJUSTMENT
310 :  D1=2:IF INT(Y/4)*4=Y THEN D1=1
320 :  IF D>91-D1 THEN D=D+D1
330 :  M=INT(D/30.57) : REM MONTH
340 :  D=D-INT(30.57*M) : REM DAY
350 :  IF M>12 THEN M=1:Y=Y+1
360 :  Y=Y-1900
370 :  REM OUTPUT TO DISPLAY
380 :  Z=D:GOSUB 500:D$=Z$+"/"
390 :  Z=M:GOSUB 500:D$=D$+Z$+"/"
400 :  Z=Y:GOSUB 500:D$=D$+Z$
410 :  PRINT D$,:
420 :  X=X+P : REM INTERVAL ADDED
430 NEXT
440 PRINT:PRINT:PRINT,CHR$(5) " ANOTHER
GO? Y OR N" CHR$(31):
450 GET G$: IF G$<>"Y" AND G$<>"N" THEN 4
50
460 IF G$="Y" THEN PRINT:RUN
470 PRINT CHR$(147) "BYE FOR NOW":END
500 REM FORMAT SUBROUTINE
510 Z$=MID$(STR$(Z),2):IF LEN(Z$)<2 THEN
 Z$="0"+Z$
520 RETURN

READY.
```

CHAPTER 8
Greatest Common Divisor

If A and B are integers (whole numbers) then a *common divisor* (or *common factor*) of A and B is an integer which divides both numbers. And, the *greatest* common divisor (or *highest* common factor) of A and B is the largest such integer.

For instance, 3 is a common divisor of 12 and 18. But 6 is the greatest common divisor of 12 and 18.

Calculating the greatest common divisor of two numbers is not particularly complicated. Especially for a computer. The method employed serves as a good illustration of a computational algorithm.

The *Euclidean algorithm* is the most well-known and oldest (third century B.C.) method of computing the greatest common divisor. If you want to find the greatest common divisor of A and B then the procedure is as follows.

1. Rename A and B (if necessary) so that A is greater than B.
2. Divide A by B and find the remainder R_1.

$$R_1 = A - B * INT(A/B)$$

Notice that every number that divides A and B also divides R_1. And, conversely, every common divisor of B and R_1 is also a divisor of A. It follows that the common divisors of A and B are the same as those of B and R_1. Thus the greatest common divisor of A and B equals the greatest common divisor of B and R_1.

3. Now divide B by R_1 and find the remainder R_2.

$$R_2 = B - R_1 * INT(B/R_1)$$

The remarks made above between the numbers B, R_1 also apply to R_1, R_2. Thus the greatest common divisor of A and B equals the greatest common divisor of R_1 and R_2.

4. Next, divide R_2 by R_1 to get a remainder R_3.

$$R_3 = R_1 - R_2 * INT(R_1/R_2)$$

The process is continued in this manner until the remainder is zero. Notice that the remainders are decreasing on each occasion and so reaches zero after a certain number of steps.

$$R_1 = A - B * INT(A/B) \qquad (0 <= R_1 < B)$$
$$R_2 = B - R_1 * INT(B/R_1) \qquad (0 <= R_2 < R_1)$$
$$R_3 = R_1 - R_2 * INT(R_1/R_2) \qquad (0 <= R_3 < R_2)$$
$$\vdots \qquad \vdots$$
$$R_{N-1} = R_{N-3} - R_{N-2} * INT(R_{N-3}/R_{N-2}) \qquad (0 <= R_{N-1} < R_{N-2})$$
$$R_N = R_{n-2} - R_{N-1} * INT(R_{N-2}/R_{N-1}) \qquad (R_N = 0)$$

When the remainder reaches 0 we see that the previous remainder R_{N-1} is the greatest common divisor of R_{N-1} and R_{N-2}. Arguing in this way we see that R_{N-1} is the greatest common divisor of A and B.

The process outlined above is easily computerised. A program doing this is given below.

```
10 REM GREATEST COMMON DIVISOR
20 PRINT CHR$(147) "        GREATEST COM
MON DIVISOR" CHR$(17)
30 PRINT "THIS PROGRAM CALCULATES THE GR
EATEST"
40 PRINT "COMMON DIVISOR OF TWO INTEGERS
 USING THEEUCLIDEAN ALGORITHM." CHR$(17)
50 PRINT "ENTER THE TWO INTEGERS." CHR$(
17)
60 INPUT " FIRST INTEGER ";A:PRINT
70 IF A<1 OR INT(A)<>A THEN PRINT "TRY A
GAIN.":GOTO 60
80 INPUT "SECOND INTEGER ";B:PRINT
90 IF B<1 OR INT(B)<>B THEN PRINT "TRY A
GAIN.":GOTO 80
100 IF A<B THEN C=A:A=B:B=C
110 REM THE EUCLIDEAN ALGORITHM
120 R=A:S=B
130 T=R-S*INT(R/S)
140 IF T<>0 THEN R=S:S=T:GOTO 130
150 PRINT "THE GREATEST COMMON DIVISOR I
S:":PRINT,,CHR$(158);S;CHR$(17)
160 PRINT CHR$(154) "THE LEAST COMMON MU
LTIPLE IS:":PRINT,,CHR$(158);A*B/S
170 PRINT:PRINT CHR$(154),"ANOTHER GO? Y
 OR N"
180 GET G$:IF G$<>"Y" AND G$<>"N" THEN 1
80
190 IF G$="Y" THEN RUN
```

```
200 PRINT CHR$(147) "BYE FOR NOW":END
```

READY.

The program also calculates the least common multiple of A and B. The least common multiple of two numbers A and B is the smallest number which is divisible by both A and B. The value of the least common multiple of A and B is given by

A*B/(greatest common divisor)

If the greatest common divisor of A and B is D then it is possible to write D as a combination of A and B:

D = S*A + T*B

where S and T are integers. The values of S and T can be found by working backwards with the Euclidean algorithm. Modify the greatest common divisor program so that it also computes S and T.

CHAPTER 9
Primes

A *prime* number is an integer (greater than 1) that is not divisible by any positive integer other than 1 and itself (of course, by divisibility we mean exact divisibility). The numbers 2, 3, 5, 7 and 11 are prime, but 4, 6, 8, 9 and 10 are not. Non-prime numbers are called *composite*.

There is no general formula for prime numbers, but Euclid showed (in about 300 B.C.) that there are infinitely many primes. We also know that primes occur less frequently among large numbers.

The testing of primes is of considerable interest. Attention has arisen recently because of cryptography.

A simple and straightforward method of determining if a number N is prime is called the Sieve of Erastosthenes. (Erastosthenes of Cyrene was a Greek mathematician, 276-196 B.C., who also calculated the circumference of the Earth.) The idea is to write down all the integers from 1 to N. Then leave 2 and strike out all even numbers after 2. The next number after 2 which has not been struck out is prime. This is 3. Now strike out every third number after 3. The next number left after 3 is 5 which must be prime. Now strike out every fifth number after 5. This process is continued. What remains are the primes between 1 and N.

The table below shows the result of a sieve on the numbers up to 100. The multiples of 2 are crossed out by /, the multiples of 3 by —, 5 by \ and 7 by |.

	2	3	4̸	5	6̸	7	8̸	9̸	10̸
11	12̸	13	14̸	15̸	16̸	17	18̸	19	20̸
21̵	22̸	23	24̸	25̸	26̸	27̵	28̸	29	30̵
31	32̸	33̵	34̸	35̸	36̵	37	38̸	39̵	40̸
41	42̸	43	44̸	45̸	46̸	47	48̸	49̸	50̸
51̵	52̸	53	54̸	55̸	56̸	57̵	58̸	59	60̸
61	62̸	63̵	64̸	65̸	66̵	67	68̸	69̵	70̸
71	72̸	73	74̸	75̸	76̸	77̸	78̵	79	80̸
81̵	82̸	83	84̸	85̸	86̸	87̵	88̸	89	90̵
91	92̸	93̵	94̸	95̸	96̸	97	98̸	99̵	100̸

The following program uses the sieve of Erastosthenes to produce a list of prime numbers less than some given number N. This number is INPUT at the start.

Mathematics on the Commodore 64

```
10 REM SIEVE OF ERASTOSTHENES
20 PRINT CHR$(147) "          SIEVE OF ERA
STOSTHENES" CHR$(17)
30 PRINT "THIS PROGRAM WILL CALCULATE TH
E PRIME"
40 PRINT "NUMBERS UPTO SOME GIVEN NUMBER
." CHR$(17)
50 INPUT "UP TO WHAT NUMBER ";N:PRINT
60 IF N<4 OR INT(N)<>N OR N>15000 THEN P
RINT "BE REASONABLE.":GOTO 50
70 DIM A%(N):PRINT CHR$(158) "PRIMES FRO
M 2 TO";N:PRINT
80 FOR I=2 TO N
90 IF A%(I)=1 THEN 130
100 IF 39-POS(0)<LEN(STR$(I)) THEN PRINT
110 PRINT I;
120 FOR J=I TO N STEP I:A%(J)=1:NEXT
130 NEXT:PRINT
140 PRINT:PRINT CHR$(154),"ANOTHER GO? Y
 OR N"
150 GET G$:IF G$<>"Y" AND G$<>"N" THEN 1
50
160 IF G$="Y" THEN RUN
170 PRINT CHR$(147) "BYE FOR NOW":END

READY.
```

Note that for large numbers the program starts off slowly; but soon starts printing primes very fast. It takes about 7 minutes and 21 seconds to print the primes between 2 and 15000.

The sieve of Erastosthenes is conceptually easy. It is useful if you want a list of prime numbers. But it is not a very practical method of testing whether a number is prime.

A simple way to check if a number N is prime is to check whether it is divisible by the numbers smaller than N, step by step. The following program illustrates this method. A clock is included to indicate how long it takes to test the number for primality.

```
10 REM PRIMES VERSION 1
20 PRINT CHR$(147) "   INEFFICIENT PRIME
 TESTER VERSION 1" CHR$(17)
30 INPUT "NUMBER TO BE TESTED ";N:PRINT
40 IF N<4 OR N<>INT(N) THEN PRINT "BE RE
ASONABLE":GOTO 30
```

Chapter 9 Primes

```
50 A$="":TI$="000000":X=N-1
60 FOR I=2 TO X
70 T=N/I:IF INT(T) = T THEN A$="NOT ":I=X
80 NEXT
90 PRINT "THE NUMBER IS " A$ "PRIME."
100 PRINT:PRINT "TIME TAKEN TO TEST NUMBER":INT(TI/60+0.5) "SECONDS"
110 PRINT:PRINT CHR$(154),"ANOTHER GO? Y OR N"
120 GET G$:IF G$<>"Y" AND G$<>"N" THEN 120
130 IF G$="Y" THEN RUN
140 PRINT CHR$(147) "BYE FOR NOW":END

READY.
```

The program works but it really is inefficient and slow. For instance to test the primality of 9001 takes about 113 seconds. A little thought will produce enormous benefits.

For a start we needn't bother using all the numbers between 2 and N. We need only use the numbers between 2 and INT(SQR(N)). Because if M is an integer which divides N and is greater than INT(SQR(N)) then N/M is an integer that divides N and is smaller than INT(SQR(N)). This simple addition is included in version two below.

```
10 REM PRIMES VERSION 2
20 PRINT CHR$(147) "     INEFFICIENT PRIME TESTER VERSION 2" CHR$(17)
30 INPUT "NUMBER TO BE TESTED ";N:PRINT
40 IF N<4 OR N<>INT(N) THEN PRINT "BE REASONABLE":GOTO 30
50 A$="":TI$="000000":X=SQR(N)
60 FOR I=2 TO X
70 T=N/I:IF INT(T) = T THEN A$="NOT ":I=X
80 NEXT
90 PRINT "THE NUMBER IS " A$ "PRIME."
100 PRINT:PRINT "TIME TAKEN TO TEST NUMBER":INT(TI/60+0.5) "SECONDS"
110 PRINT:PRINT CHR$(154),"ANOTHER GO? Y OR N"
120 GET G$:IF G$<>"Y" AND G$<>"N" THEN 120
```

```
130 IF G$="Y" THEN RUN
140 PRINT CHR$(147) "BYE FOR NOW":END
```

READY.

This version works considerably faster. Testing the number 9001 now takes just over 1 second. A larger number, such as 987654323 takes about 387 seconds to test. Don't attempt to test such a large number with version 1 — unless you enjoy looking mindlessly at your television screen.

Remark. The addition incorporated in Prime Version 2 could also be incorporated in the program Erastosthenes to produce Erastosthenes 2. With this addition it takes 6 minutes and 21 seconds to print out the primes from 2 to 15000.

```
10 REM SIEVE OF ERASTOSTHENES : WITH ADD
ITIONS
20 PRINT CHR$(147) "        SIEVE OF ERA
STOSTHENES" CHR$(17)
30 PRINT "THIS PROGRAM WILL CALCULATE TH
E PRIME"
40 PRINT "NUMBERS UPTO SOME GIVEN NUMBER
." CHR$(17)
50 INPUT "UP TO WHAT NUMBER ";N:PRINT
60 IF N<4 OR INT(N)<>N OR N>15000 THEN P
RINT "BE REASONABLE.":GOTO 50
70 DIM A%(N):PRINT CHR$(158) "PRIMES FRO
M 2 TO";N:PRINT:S=INT(SQR(N)+1)
80 FOR I=2 TO N
90 IF A%(I)=1 THEN 130
100 IF 39-POS(0)<LEN(STR$(I)) THEN PRINT
110 PRINT I;
120 IF I<=S THEN FOR J=I TO N STEP I:A%(
J)=1:NEXT
130 NEXT:PRINT
140 PRINT:PRINT CHR$(154),"ANOTHER GO? Y
 OR N"
150 GET G$:IF G$<>"Y" AND G$<>"N" THEN 1
50
160 IF G$="Y" THEN RUN
170 PRINT CHR$(147) "BYE FOR NOW":END
```

READY.

We can further improve the program Primes Version 2 by taking a tip from the sieve of Erastosthenes. We can miss out all even numbers bigger than 2 and miss out every third number after 3. These suggestions have been added to the third version of the program.

```
10 REM PRIMES VERSION 3
20 PRINT CHR$(147),"PRIME TESTER VERSION
    3" CHR$(17)
30 INPUT "NUMBER TO BE TESTED ";N:PRINT
40 IF N<4 OR N<>INT(N) THEN PRINT "BE RE
ASONABLE":GOTO 30
50 A$="":TI$="000000":X=SQR(N)
52 IF INT(N/2) = N/2 THEN A$="NOT ":GOTO
    90
53 IF INT(N/3) = N/3 THEN A$="NOT ":GOTO
    90
60 FOR I=5 TO X STEP 6
70 T=N/I:IF INT(T) = T THEN A$="NOT ":I=
X
75 T=N/(I+2):IF INT(T) = T THEN A$="NOT
":I=X
80 NEXT
90 PRINT "THE NUMBER IS " A$ "PRIME."
100 PRINT:PRINT "TIME TAKEN TO TEST NUMB
ER":INT(TI/60+0.5) "SECONDS"
110 PRINT:PRINT CHR$(154),"ANOTHER GO? Y
 OR N"
120 GET G$:IF G$<>"Y" AND G$<>"N" THEN 1
20
130 IF G$="Y" THEN RUN
140 PRINT CHR$(147) "BYE FOR NOW":END
```

READY.

This speeds up the program a little. Testing the prime 987654323 now takes about 133 seconds which is quite reasonable.

A composite number may be written as a product of prime numbers. For instance

6 = 3 * 2
24 = 3 * 2 * 2 * 2
81018001 = 9001 * 9001

and so on. The various primes that occur are called the *factors* of the number. By adding a few extra lines to version 3 we can have all the factors of a number displayed.

The next program prints out the factors of a number.

93

```
10 REM PRIME FACTORS
20 GOTO 70 : REM NEXT BIT IS SUBROUTINE
 FOR FACTORS
30 Y=N/T:IF 38-POS(0)<LEN(STR$(Y)) THEN
 PRINT
40 PRINT Y::N=T:T=N/Y:IF T=INT(T) THEN 3
0
50 S=SQR(N)+1:RETURN
60 REM THE MAIN PART
70 PRINT CHR$(147),CHR$(154) "    PRIME F
ACTORS" CHR$(17)
80 INPUT "NUMBER TO BE TESTED ";N:PRINT
90 IF N<4 OR N<>INT(N) THEN PRINT "BE RE
ASONABLE":GOTO 80
100 PRINT CHR$(5) "THE FACTORS OF";N;"AR
E" CHR$(17)
110 TI$="000000":X=SQR(N):S=X+1
120 T=N/2:IF INT(T) = T THEN GOSUB 30
130 T=N/3:IF INT(T) = T THEN GOSUB 30
140 FOR I=5 TO X STEP 6
150 T=N/I:IF INT(T) = T THEN GOSUB 30
160 T=N/(I+2):IF INT(T) = T THEN GOSUB 3
0
170 IF I>S THEN I=X
180 NEXT
190 IF 38-POS(0)<LEN(STR$(N)) THEN PRINT
200 IF N>1 THEN PRINT N;
210 PRINT:PRINT:PRINT CHR$(158) "THAT'S
ALL"
220 PRINT:PRINT "TIME TO FIND FACTORS WA
S";INT(TI/60+0.5) "SECONDS"
230 PRINT:PRINT CHR$(154),"ANOTHER GO? Y
 OR N"
240 GET G$:IF G$<>"Y" AND G$<>"N" THEN 2
40
250 IF G$="Y" THEN RUN
260 PRINT CHR$(147) "BYE FOR NOW":END

READY.
```

If you enter a very large number into the program then the Commodore 64 will round off the number and find the factors of that number.

Chapter 9 Primes

Large primes

Finding large prime numbers is a pastime for some. The largest known primes are usually Mersenne primes. Numbers of the form

$$2^P - 1$$

are called *Mersenne numbers* because of the French monk, Father Marin Mersennes who made some suggestions concerning the primality of such numbers.

Many early writers believed that the Mersenne numbers are prime if the exponent M is prime. For the first few cases this is indeed true.

$$2^2 - 1 = 3$$
$$2^3 - 1 = 7$$
$$2^5 - 1 = 31$$
$$2^7 - 1 = 127$$

But for P = 11 the number is not prime.

$$2^{11} - 1 = 2047 = 23 * 89$$

For at least the last 100 years the world's largest known prime has always been a Mersenne prime (except for a short period in 1951). Prior to January 1983, the largest known prime was

$$2^{44497} - 1$$

a Mersenne prime discovered by David Slowinski in April 1979. This was the 27th Mersenne prime known. In about January 1983 Slowinski found the much larger prime

$$2^{86243} - 1$$

which is a Mersenne prime containing 25962 decimal digits. (It would fill several pages of this book to write out the number completely.) The computer that Slowinski used was a Cray−1. This is an amazingly fast machine, even so it took 1 hour 36 minutes and 22 seconds of computer time to perform the test.

An arbitrary number of the size of $2^{286243} - 1$ would be impossible to test for primality using any of the programs in the previous section. But there are special techniques for Mersenne numbers. These were developed during the last 100 years. To test a number of the form $N = 2^P - 1$ for primality one defines a sequence as follows:

$$U_1 = 4$$
$$U_2 = U_1*U_1 - N*INT((U_1*U_1-2)/N)$$
$$\vdots \quad \vdots$$
$$U_{p-1} = U_{p-2}*U_{p-2} - 2 - N*INT((U_{p-2}*U_{p-2}-2)/N)$$

Then N is prime if and only if $U_{p-1} = 0$. This means that in order to test for N being prime we need to perform approximately P simple calculations. Performing 86243 operations would be quite simple for the Commodore 64 if the numbers involved were small. But here one is dealing with numbers that have 86243 binary digits and the Commodore 64 stores 32 binary digits. There are methods available to get around this problem but we won't go into the details here. See the next chapter.

Probabilistic primality testing

Given unlimited time we could test for the primality of an integer by trial division. But time is limited, even for a computer.

In the last few years a new test has been devised which is based on an old theorem of Pierre Fermat who lived during the 17th century. Fermat showed that if P is a prime number and B is some other number between 1 and $P - 1$ then the number $B^{P-1} - 1$ is divisible by P. For instance, let $P = 11$ and $B = 2$ then $2^{11-1} - 1$ is 1023 which is indeed divisible by 11.

Although Fermat proved his theorem for all values of B, the Chinese mathematician Pomerance (5th Century B.C.) knew the theorem in the case $B = 2$. In addition he believed, incorrectly, that the converse is true. In other words he believed that if $2^{P-1} - 1$ is divisible by P then P is a prime number. However the number 341 divides $2^{340} - 1$ and 341 is not prime. We call such a number a *pseudoprime to the base 2*. In general a composite number P that divides $B^{P-1} - 1$ is called a *pseudoprime to the base B*. Most numbers that appear to be pseudoprimes are in fact genuine primes. And this is what the test is based on.

```
10 REM PROBABILISTIC PRIMALITY TEST
20 PRINT CHR$(147) CHR$(154) "        PROB
ABALISTIC PRIMALITY TEST" CHR$(17)
30 INPUT "NUMBER TO BE TESTED ";N
40 IF N<4 OR N<>INT(N) THEN PRINT "BE RE
ASONABLE":PRINT:GOTO 30
50 REM FACTOR N-1 = (2^T)*X
60 T=0:X=N-1
70 D=X/2:IF INT(D)=D THEN T=T+1:X=D:GOTO
   70
80 REM SELECT BASE B
```

Chapter 9 Primes

```
 90 B=RND(-TI):B=INT(RND(1)*(N-3.0001)+
2)
100 PRINT:PRINT CHR$(5) "TEST USING BASE
";B;CHR$(17)
110 REM RAISE B TO POWER X
120 P=1
130 IF X=0 THEN 170
140 D=X/2:IF D<>INT(D) THEN P=P*B:P=P-N*
INT(P/N)
150 B=B*B:B=B-N*INT(B/N):X=INT(D):GOTO 1
30
160 REM CHECK B^X
170 IF P=1 OR P=N-1 THEN 200
180 IF T<2 THEN PRINT "NUMBER IS NOT PRI
ME":GOTO 250
190 P=P*P:P=P-N*INT(P/N):T=T-1:GOTO 170
200 PRINT "NUMBER IS PROBABLY PRIME" CHR
$(17)
210 PRINT CHR$(158) "DO YOU WANT TO TEST
 THE NUMBER AGAIN"
220 PRINT "WITH ANOTHER BASE? Y OR N"
230 GET G$:IF G$<>"Y" AND G$<>"N" THEN 2
30
240 IF G$="Y" THEN 60
250 PRINT:PRINT CHR$(154),"ANOTHER GO? Y
 OR N"
260 GET G$:IF G$<>"Y" AND G$<>"N" THEN 2
60
270 IF G$="Y" THEN RUN
280 PRINT CHR$(147) "BYE FOR NOW":END

READY.
```

Try numbers such as 341 and 561, neither of which are prime. If you find that they are a pseudoprime to some base then try another base.

Note. The program is fast but doesn't work very well with large numbers because of rounding off in the multiplications involved.

CHAPTER 10
Odds and Ends

Pythagorean triplets

Recall the theorem of Pythagoras which has already been mentioned in this book. It states that in a right angled triangle with sides X, Y and Z, where Z is the hypotenuse (the longest side), the following relation holds.

$X^2 + Y^2 = Z^2$

The classic example that is readily recalled is the 3, 4, 5 right triangle.

$3^2 + 4^2 = 5^2$

Another example is the 5, 12, 13 right triangle.

$5^2 + 12^2 = 13^2$

Of course there are infinitely many different examples of right angled triangles. But, how many examples are there in which X, Y and Z are integers? The answer is that there are infinitely many. Such numbers X, Y and Z are called *Pythagorean triplets*.

How can we produce a list of some Pythagorean triplets? Naturally such a list should not contain triplets that are products of another. For instance, since 3, 4, 5 is a Pythagorean triplet so is 6, 8, 10. Pythagorean triplets which have no common factor are called *primitive*. Thus 3, 4, 5 is a primitive pythagorean triplet while 6, 8, 10 is not primitive.

Producing Pythagorean triplets could be time consuming were it not for some mathematicians who managed to produce an elegant method of generating primitive Pythagorean triplets. The technique is as follows.

1. Choose two positive integers A and B so that:

   (a) A is greater than B.
   (b) A + B is odd (thus one of the numbers is odd and the other is even)
   (c) A and B have no common divisor except 1.

2. Calculate

$$X = A^2 - B^2$$
$$Y = 2*A*B$$
$$Z = A^2 + B^2$$

The numbers X, Y, Z are a primitive Pythagorean triplet. Conversely, any primite Pythagorean triplet can be formed in this way. This technique is used in the program Pythagorean Triplets to produce a list of pythagorean triplets. The process starts with A = 2. The program then finds all possible values of B satisfying the required conditions. The value of A is increased and the process repeated. Twenty triplets are printed out at a time. You may continue for as long as you like. The starting value of A may be changed if desired.

```
10 REM PYTHAGOREAN TRIPLETS
20 PRINT CHR$(147),CHR$(154) "PYTHAGOREA
N TRIPLETS" CHR$(17)
30 PRINT "THIS PROGRAM PRINTS OUT SOME P
RIMITIVE"
40 PRINT "PYTHAGOREAN TRIPLETS." CHR$(17
)
50 PRINT,"   PRESS Y TO START":K=0:A=2:B=
3
60 GET G$:IF G$<>"Y" THEN 60
70 PRINT CHR$(147),CHR$(154) "PYTHAGOREA
N TRIPLETS" CHR$(17)
80 PRINT "COUNT     ** X **    ** Y **    **
 Z **" CHR$(158)
90 B=B-2:IF B<1 THEN A=A+1:B=A-1
100 A1=A:B1=B
110 N=INT(A1/B1):A2=A1-N*B1
120 IF A2<>0 THEN A1=B1:B1=A2:GOTO 110
130 IF B1<>1 THEN 90
140 K=K+1:PRINT K,A*A-B*B,2*A*B,A*A+B*B
150 IF INT(K/20)<>K/20 THEN 90
160 PRINT:PRINT CHR$(154)," CONTINUE?   Y
 OR N";
170 GET G$:IF G$<>"Y" AND G$<>"N" THEN 1
70
180 IF G$="Y" THEN 70
190 PRINT CHR$(147) "BYE FOR NOW":END

READY.
```

Chapter 10 Odds and Ends

Multi-precision powers

The Commodore 64 keeps numbers with 9 significant digits accurately. Thus integers less than 999,999,999 are stored accurately. Larger numbers are rounded off. How can we calculate large powers of numbers accurately? For instance, what is 2 to the power 130? The answer is to use multi-precision arithmetic.

One way to produce multi-precision arithmetic is to store the digits of a number in an array M(I). With M(0) storing the last 4 digits, M(1) storing the previous 4 and so on. Thus 987654 would be stored as:

M(1) = 98, M(0) = 7654

The original number is recovered by converting the array elements into strings and printing each in turn. Of course, it may also be recovered by the formula:

M = M(1)*10^4 + M(0)

Suppose that we want to multiply M = 987654 by N = 23456. First we would write these numbers into arrays M(I) and N(I).

M(1) = 98, M(0) = 7654
N(1) = 2, N(0) = 3456

Now form the following

C(0) = M(0)*N(0)
C(1) = M(1)*N(0) + M(0)*N(1)
C(2) = M(1)*N(1)

The result becomes:

C(0) = 26452224
C(1) = 353996
C(2) = 196

We then calculate the product M*N by the following

196*10^8 + 353996*10^4 + 26452224
= 19600000000 + 3539960000 + 26452224
= 23166412224

Of course the calculation is not done as shown above — the Commodore 64 would simply round off the numbers. What we do is strip

101

off the digits from the left of C(0) so that only 4 are left. We add the digits stripped off to C(1).

$$C(0) \rightarrow 2224$$
$$C(1) \rightarrow 353996 + 2645 = 356641$$

Next, strip off the digits from the left of C(1) so that only 4 are left. Add the digits stripped off to C(2).

$$C(1) \rightarrow 6641$$
$$C(2) \rightarrow 196 + 35 = 231$$

We therefore obtain the following:

$$C(2) = 231, C(1) = 6641, C(0) = 2224$$

from which we read immediately that

$$987654*23456 = 23166412224$$

This process is quite general. For other larger numbers there may be a value for the array element M(2), M(3) etc. In such a situation we would form the following.

$$C(0) = M(0)*N(0)$$
$$C(1) = M(1)*N(0) + M(0)*N(1)$$
$$C(2) = M(2)*N(0) + M(1)*N(1) + M(0)*N(2)$$
etc.

After this the stripping process is performed when finally the answer can be printed out.

The program Multi-Precision Powers illustrates how the process just described may be used to calculate accurately powers of numbers. The program continues until a 40 digit number is reached. If desired you can have even higher degrees of accuracy by changing the value of X in line 80. The value of X + 1 times 4 is the degree of precision.

```
10 REM MULTI-PRECISION POWERS
20 PRINT CHR$(147),CHR$(154) "MULTI-PREC
ISION POWERS" CHR$(17)
30 PRINT "THIS PROGRAM PRINTS OUT POWERS
 OF AN"
40 PRINT "INTEGER ACCURATE TO 40 DIGITS.
" CHR$(17)
50 PRINT "ENTER NUMBER WHOSE POWERS ARE
REQUIRED. "
60 INPUT "NUMBER ";N:PRINT
70 IF N<2 OR N>999999999 OR N<>INT(N) TH
EN PRINT "ERROR - TRY AGAIN":GOTO 60
```

Chapter 10 Odds and Ends

```
80 T=10000:X=9:Y=X+2:DIM M(X),N(X),L(Y):
K=1:N(0)=N:M(0)=N
85 REM CHANGE X FOR OTHER DEGREES OF PRE
CISION
90 FOR I=0 TO 2
100 IF M(I)>=T THEN Q=INT(M(I)/T):M(I)=M
(I)-Q*T:M(I+1)=M(I+1)+Q
110 N(I)=M(I)
120 NEXT
130 PRINT CHR$(147) "MULTI-PRECISION POW
ERS OF";N;CHR$(17)
140 K=K+1
150 FOR I=0 TO Y:L(I)=0:NEXT
160 FOR J=0 TO 2:FOR I=0 TO X
170 L(I+J)=L(I+J)+N(I)*M(J)
180 NEXT:NEXT
190 FOR I=0 TO X:N(I)=L(I):NEXT
200 FOR I=0 TO X
210 IF N(I)>=T THEN Q=INT(N(I)/T):N(I)=N
(I)-Q*T:N(I+1)=N(I+1)+Q
220 NEXT
230 FOR I=0 TO X
240 IF N(I)>0 THEN L=I
250 NEXT
260 PRINT CHR$(154);N;"^";K;CHR$(158)
270 FOR I=L TO 0 STEP -1
280 A$=MID$(STR$(N(I)),2)
290 IF I<L AND LEN(A$)<>4 THEN A$="0"+A$
:GOTO 290
300 PRINT A$;:NEXT:PRINT
310 IF (N(X)*T+N(X-1))*N>=T*T THEN 360
320 IF INT(K/10)<>K/10 THEN 140
330 PRINT:PRINT CHR$(154)," PRESS Y TO C
ONTINUE";
340 GET G$:IF G$<>"Y" THEN 340
350 GOTO 130
360 PRINT:PRINT CHR$(154),"ANOTHER GO? Y
 OR N";
370 GET G$:IF G$<>"Y" AND G$<>"N" THEN 3
70
380 IF G$="Y" THEN RUN
390 PRINT CHR$(147) "BYE FOR NOW":END

READY.
```

CHAPTER 11
Matrices

Matrices are rectangular arrays of numbers, the position and value of each number being important. A matrix is usually, but not always, shown in brackets thus:

$$\begin{bmatrix} 2 & -3 & 6 \\ -4 & 4 & 5 \end{bmatrix}$$

The size of a matrix is given by the number of rows and columns. We say that a matrix is an M by N matrix if it has M rows and N columns. Thus the example above is a 2 by 3 matrix. Here are some examples of matrices

$$\begin{bmatrix} 8 & 12 & 0 \\ 1 & 1 & 8 \\ 8 & 12 & 1 \end{bmatrix} \quad \text{A 3 by 3 matrix}$$

$$\begin{bmatrix} 9 & 12 \\ 0 & 1 \\ 8 & -2 \end{bmatrix} \quad \text{A 3 by 2 matrix}$$

$$\begin{bmatrix} 9 & 12 \\ 3 & 2 \end{bmatrix} \quad \text{A 2 by 2 matrix}$$

If the number of rows in a matrix equals the number of columns in the matrix then we say that it is a square matrix.

Matrices are a useful concept for holding information in a concise way. For instance, suppose that you have calculated the amount spent in your household on three types of fuel each quarter during a year. You could display the information in a tables as follows:

Quarter	Gas	Electric	Solid fuel
1	30	28	5
2	27	19	4
3	25	15	0
4	32	27	3

105

The data can be stored in a 4 by 3 matrix.

$$\begin{bmatrix} 30 & 28 & 5 \\ 27 & 19 & 4 \\ 25 & 15 & 0 \\ 32 & 27 & 3 \end{bmatrix}$$

You can store matrices in your Commodore 64 by using two-dimensional arrays such as A(I,J). For instance, the matrix of fuel expenditure could be stored by making the following assignments.

A(1,1) = 30 A(1,2) = 28 A(1,3) = 5
A(2,1) = 27 A(2,2) = 19 A(2,3) = 4
A(3,1) = 25 A(3,2) = 15 A(3,3) = 0
A(4,1) = 32 A(4,2) = 27 A(4,3) = 3

The information would, of course, be entered in by using READ statements.

FOR I = 1 TO 4:FOR J = 1 TO 3:READ A(I,J):NEXT:NEXT
DATA 30,28,5,27,19,4,25,15,0,32,27,3

The following short program illustrates the data being read in and then displayed.

```
10 PRINT CHR$(147)
20 FOR I = 1 TO4:FOR J = 1 TO 3:READ A(I,J):NEXT:NEXT
30 FOR I = 1 TO 4
40 FOR J = 1 TO 3:PRINT A(I,J);:NEXT
50 PRINT
60 NEXT
100 DATA 30,28,5,27,19,4,25,15,0,32,27,3
```

Note that to use the memory space of the Commodore 64 more efficiently we should have started the I and J counters at 0 instead of 1 since arrays begin at 0. However, conceptually the above is easier.

Also note that if the number of rows or columns in a matrix exceeds 10 then a DIM statement is required by the Commodore 64.

Adding matrices

Two matrices may be added together provided that they are of the same size. In other words two matrices may be added together provided that they both have the same number of rows and the same number of columns. This addition is performed term-by-term of the corresponding terms in the corresponding positions. For example

$$\begin{bmatrix} 1 & 5 & 4 \\ -2 & 3 & 1 \end{bmatrix} + \begin{bmatrix} 2 & 0 & 5 \\ 3 & 2 & 1 \end{bmatrix}$$

$$= \begin{bmatrix} 1+2 & 5+0 & 4+5 \\ -2+3 & 3+2 & 1+1 \end{bmatrix}$$

$$= \begin{bmatrix} 3 & 5 & 9 \\ 1 & 5 & 2 \end{bmatrix}$$

Why add matrices?

An example of the expenditure on fuel for the four quarters of a year was given earlier on. Suppose that during the next year the corresponding matrix is as follows:

$$\begin{bmatrix} 34 & 27 & 4 \\ 30 & 20 & 3 \\ 20 & 16 & 1 \\ 35 & 29 & 5 \end{bmatrix}$$

By adding the two matrices together we can obtain the amount spent on the three fuels during a particular quarter in the two years. The result is the following matrix:

$$\begin{bmatrix} 64 & 55 & 9 \\ 57 & 39 & 7 \\ 45 & 31 & 1 \\ 67 & 56 & 8 \end{bmatrix}$$

Your Commodore 64 can add matrices, or two-dimensional arrays quite easily. Suppose that we have two arrays A(I,J) and B(I,J) where in each case I ranges from 1 to 4 and J ranges from 1 to 3. Addition of these two arrays will produce a third array C(I,J) in the following way:

   FOR I = 1 TO 4:FOR J = 1 TO 3
   C(I,J) = A(I,J) + B(I,J)
   NEXT:NEXT

Subtraction of matrices uses the same rules as additions of matrices.

Matrix multiplication

Matrix multiplication is a quite subtle concept. In order that two matrices may be multiplied it is required that the number of columns in the first matrix is equal to the number of rows in the second matrix. Thus a 4 by 3 matrix can be multiplied with a 3 by 2 matrix. But a 4 by 3 matrix cannot be multiplied with a 2 by 3 matrix.

The product of an M by N matrix with an N by P matrix will be an M by P matric. The actual process is best illustrated with an example, the explanation being given afterwards. Suppose we want to multiply the following matrices:

$$\begin{bmatrix} 2 & -3 & 6 \\ -4 & 4 & 5 \end{bmatrix}$$ First matrix, a 2 by 3 matrix

$$\begin{bmatrix} 9 & 12 \\ 0 & 1 \\ 8 & -2 \end{bmatrix}$$ Second matrix, a 3 by 2 matrix

The number of columns in the first matrix equals the number of rows in the second and so we may multiply the first by the second.

$$\begin{bmatrix} 2 & -3 & 6 \\ -4 & 4 & 5 \end{bmatrix} * \begin{bmatrix} 9 & 12 \\ 0 & 1 \\ 8 & -2 \end{bmatrix}$$

$$= \begin{bmatrix} 2*9 + -3*0 + 6*8 & 2*12 + -3*1 + 6*-2 \\ -4*9 + 4*0 + 5*8 & -4*12 + 4*1 + 4*-2 \end{bmatrix}$$

$$= \begin{bmatrix} 66 & 9 \\ 4 & -52 \end{bmatrix}$$

To perform the multiplication first look at each row of the first matrix and each column of the second matrix. The number of elements in each of these is the same. For each of these rows and columns we multiply the first of each of these together, then the second, and so on then finally add these products together. This produces a number for the product matrix.

On your Commodore 64 you can form the product of the M by N matrix A(I,J) with the N by P matrix B(I,J) in the following way. The result is an M by P matrix C(I,J).

```
FOR I=1 TO M
FOR J=1 TO P
C(I,J) = 0
FOR K=1 TO N : C(I,J) = C(I,J) + A(I,K)*B(K,J) : NEXT
NEXT
NEXT
```

Note that if A and B are two matrices and if you can form the product A*B then you may not be able to form the product B*A. Even if the product B*A exists it is not necessarily equal to A*B. Find some simple examples!

Why multiply matrices?

Lets's look at our household fuel costs example once again. Recall the original details.

Chapter 11 Matrices

Quarter	Gas	Electric	Solid fuel
1	30	28	5
2	27	19	4
3	25	15	0
4	32	27	3

Suppose that there have been two estimates of the likely increases in the costs of these fuels.

	Estimate 1	Estimate 2
Gas	10%	5%
Electric	5%	10%
Solid fuel	10%	10%

These increases are given as decimals in the next table.

	Estimate 1	Estimate 2
Gas	0.1	0.05
Electric	0.05	0.1
Solid fuel	0.1	0.1

How much extra would you have to pay for the three fuels during each quarter? The answer depends on which estimate you use. The result may be tabulated as shown in the next table.

Quarter	Estimate 1	Estimate 2
1	30*.1 + 28*.05 + 5*.1	30*.05 + 28*.1 + 5*.1
2	27*.1 + 19*.05 + 4*.1	27*.05 + 19*.1 + 4*.1
3	25*.1 + 15*.05 + 0*.1	25*.05 + 15*.1 + 0*.1
4	32*.1 + 27*.05 + 3*.1	32*.05 + 27*.1 + 3*.1

This evaluates to give the following table.

Quarter	Estimate 1	Estimate 2
1	4.9	4.8
2	4.05	3.65
3	3.24	2.75
4	4.85	4.6

You have probably seen that the resulting matrix is the product of the matrix of expenditure with the matrix of estimates. In other words it is the following product of matrices.

$$\begin{bmatrix} 30 & 28 & 5 \\ 27 & 19 & 4 \\ 25 & 15 & 0 \\ 32 & 27 & 3 \end{bmatrix} * \begin{bmatrix} 0.1 & 0.05 \\ 0.05 & 0.1 \\ 0.1 & 0.1 \end{bmatrix}$$

Zeros and ones

Matrices that consist of zeros alone are called zero matrices. Adding a zero matrix to another matrix has no effect. Multiplying a matrix by a zero matrix produces a zero matrix. For instance:

$$\begin{bmatrix} 30 & 28 \\ 27 & 19 \\ 25 & 15 \end{bmatrix} * \begin{bmatrix} 0 & 0 \\ 0 & 0 \end{bmatrix}$$

$$= \begin{bmatrix} 30*0 + 28*0 & 30*0 + 28*0 \\ 27*0 + 19*0 & 27*0 + 19*0 \\ 25*0 + 15*0 & 25*0 + 15*0 \end{bmatrix}$$

$$= \begin{bmatrix} 0 & 0 \\ 0 & 0 \\ 0 & 0 \end{bmatrix}$$

The product of two non-zero matrices can be zero. An example is given below.

$$\begin{bmatrix} 1 & 1 \\ 1 & 1 \end{bmatrix} * \begin{bmatrix} 1 & -1 \\ -1 & 1 \end{bmatrix}$$

$$= \begin{bmatrix} 1*1 + 1*-1 & 1*-1 + 1*1 \\ 1*1 + 1*-1 & 1*-1 + 1*1 \end{bmatrix}$$

$$= \begin{bmatrix} 0 & 0 \\ 0 & 0 \end{bmatrix}$$

A square matrix that has ones down the diagonal (top left to bottom right) and zeros elsewhere is called an *identity* matrix. For example, both of the following are identity matrices.

$$\begin{bmatrix} 1 & 0 & 0 \\ 0 & 1 & 0 \\ 0 & 0 & 1 \end{bmatrix} \quad \begin{bmatrix} 1 & 0 \\ 0 & 1 \end{bmatrix}$$

A matrix multiplied by an identity matrix equals itself. In other words the identity matrix behaves much like a 1 does in ordinary number multiplication. For example:

$$\begin{bmatrix} 30 & 28 \\ 27 & 19 \\ 25 & 15 \end{bmatrix} * \begin{bmatrix} 1 & 0 \\ 0 & 1 \end{bmatrix}$$

$$= \begin{bmatrix} 30*1 + 28*0 & 30*0 + 28*1 \\ 27*1 + 19*0 & 27*0 + 19*1 \\ 25*1 + 15*0 & 25*0 + 15*1 \end{bmatrix}$$

$$= \begin{bmatrix} 30 & 28 \\ 27 & 19 \\ 25 & 15 \end{bmatrix}$$

Inverses of matrices

The inverse or reciprocal of a matrix, if it exists, has the same property as that of the inverse of an ordinary number. The inverse of the number 4 is 0.25 and

$$4 * 0.25 = 1, \quad 0.25 * 4 = 1$$

For a matrix A the inverse (if it exists) is denoted by A^{-1} and

$$A * A^{-1} = I, \quad A^{-1} * A = I$$

where I denotes an identity matrix.

Only square matrices can have inverses. In fact only certain square matrices have inverses.

Several methods exist for finding an inverse of a square matrix. There is a general step-by-step method well suited for your Commodore 64. It consists of performing so-called row operations. Roughly speaking we place an identity matrix next to the matrix whose inverse we want. The row operations are performed on both matrices simultaneously until the original is converted to an identity matrix. The matrix that was the identity is now the inverse matrix.

A program that finds inverses of matrices is given later on. Meanwhile the method is briefly described.

Suppose we want to find the inverse of the following matrix.

$$\begin{bmatrix} 2 & 1 \\ -2 & 4 \end{bmatrix}$$

Place the identity 2 by 2 matrix next to this matrix.

$$\begin{bmatrix} 2 & 1 \\ -2 & 4 \end{bmatrix} \quad \begin{bmatrix} 1 & 0 \\ 0 & 1 \end{bmatrix}$$

We want the top left entry to be a 1 and so we divide the whole first row by 2.

$$\begin{bmatrix} 1 & 0.5 \\ -2 & 4 \end{bmatrix} \quad \begin{bmatrix} 0.5 & 0 \\ 0 & 1 \end{bmatrix}$$

Next, we want the bottom left entry to be 0 and so we add twice the first row to the second row.

$$\begin{bmatrix} 1 & 0.5 \\ 0 & 5 \end{bmatrix} \quad \begin{bmatrix} 0.5 & 0 \\ 1 & 1 \end{bmatrix}$$

Now we want the bottom right entry (of the left matrix) to be 1. This is achieved by dividing the second row by 5.

$$\begin{bmatrix} 1 & 0.5 \\ 0 & 1 \end{bmatrix} \quad \begin{bmatrix} 0.5 & 0 \\ 0.2 & 0.2 \end{bmatrix}$$

To make the left matrix the identity we subtract half of the second row from the first.

$$\begin{bmatrix} 1 & 0 \\ 0 & 1 \end{bmatrix} \quad \begin{bmatrix} 0.4 & -0.1 \\ 0.2 & 0.2 \end{bmatrix}$$

The matrix on the right is the inverse of the matrix we started with. The following calculation verifies this fact.

$$\begin{bmatrix} 2 & 1 \\ -2 & 4 \end{bmatrix} * \begin{bmatrix} 0.4 & -0.1 \\ 0.2 & 0.2 \end{bmatrix}$$

$$= \begin{bmatrix} 2*.4 + 1*.2 & 2*-.1 + 1*.2 \\ -2*.4 + 4*.2 & -2*-.1 + 4*.2 \end{bmatrix}$$

$$= \begin{bmatrix} 1 & 0 \\ 0 & 1 \end{bmatrix}$$

The next program finds inverses of square matrices, that is, if an inverse exists.

```
10 REM INVERSES OF MATRICES
20 PRINT CHR$(147),CHR$(154) "INVERSES O
F MATRICES" CHR$(17)
30 PRINT "THIS PROGRAM DETERMINES THE IN
VERSE OF  AN N BY N MATRIX." CHR$(17)
40 PRINT "ENTER THE SIZE OF THE SQUARE M
ATRIX." CHR$(17)
50 INPUT "VALUE OF N ";N:PRINT
60 IF N<1 OR N<>INT(N) THEN PRINT "NONSE
NSE!! TRY AGAIN.":GOTO 50
70 PRINT "ENTER THE MATRIX TERM BY TERM.
" CHR$(17)
80 DIM A(N,N),B(N,N)
90 REM READ IN MATRIX
```

```
100 FOR I=1 TO N
110 PRINT CHR$(159);"    ROW";I;CHR$(154)
120 FOR J=1 TO N
130 PRINT "COLUMN";J;:INPUT A(I,J)
140 NEXT:PRINT:NEXT
150 REM CALCULATING
160 PRINT CHR$(147) "THE MATRIX:" CHR$(17)
170 FOR I=1 TO N:FOR J=1 TO N:PRINT A(I,J);:NEXT:PRINT:NEXT
180 PRINT:PRINT "CALCULATING ";
190 FOR I=1 TO N:B(I,I)=1:NEXT
200 X=0:GOSUB 300
210 REM ENDING
220 PRINT:PRINT CHR$(154),"ANOTHER GO? Y OR N"
230 GET G$:IF G$<>"Y" AND G$<>"N" THEN 230
240 IF G$="Y" THEN RUN
250 PRINT CHR$(147) "BYE FOR NOW":END
300 REM CALCULATING
310 X=X+1:Z=X:PRINT "*";
320 IF A(Z,X)=0 THEN Z=Z+1: IF Z<=N THEN 320
330 IF Z>N THEN PRINT:PRINT:PRINT "NO INVERSE!!":RETURN
340 IF Z<>X THEN R=1/A(Z,X):I=X:K=Z:GOSUB 500
350 IF A(X,X)<>1 THEN R=1/A(X,X):I=X:GOSUB 550
360 FOR I=1 TO N
370 IF I=X THEN I=I+1:IF I>N THEN 390
380 IF A(I,X)<>0 THEN R=-A(I,X):K=X:GOSUB 500
390 NEXT
400 IF X<N THEN 310
410 PRINT:PRINT:PRINT "THE INVERSE:" CHR$(17)
420 FOR I=1 TO N:FOR J=1 TO N:PRINT B(I,J);:NEXT:PRINT:NEXT
430 RETURN
500 REM ROW OPERATION ADD R TIMES ROW K TO ROW I
510 FOR J=1 TO N
```

```
520 A(I,J)=A(I,J)+R*A(K,J):B(I,J)=B(I,J)
+R*B(K,J)
530 NEXT
540 RETURN
550 REM ROW OPERATION MULTIPLY ROW I BY
R
560 FOR J=1 TO N
570 A(I,J)=R*A(I,J):B(I,J)=R*B(I,J)
580 NEXT
590 RETURN

READY.
```

Simultaneous equations

Matrices are useful in solving simultaneous equations. An example of 2 simultaneous equations in 2 unknowns is given below.

$$3*X + 1*Y = 7$$
$$5*X + 2*Y = 9$$

This is written in matrix form as follows.

$$\begin{bmatrix} 3 & 1 \\ 5 & 2 \end{bmatrix} * \begin{bmatrix} X \\ Y \end{bmatrix} = \begin{bmatrix} 7 \\ 9 \end{bmatrix}$$

One way to solve the problem is to find the inverse of the 2 by 2 matrix on the left and then multiply the above equation by this inverse. In the example the inverse is given by the following matrix.

$$\begin{bmatrix} 2 & -1 \\ -5 & 3 \end{bmatrix}$$

We therefore obtain the following equations.

$$\begin{bmatrix} X \\ Y \end{bmatrix}$$
$$= \begin{bmatrix} 1 & 0 \\ 0 & 1 \end{bmatrix} * \begin{bmatrix} X \\ Y \end{bmatrix}$$
$$= \begin{bmatrix} 2 & -1 \\ -5 & 3 \end{bmatrix} * \begin{bmatrix} 3 & 1 \\ 5 & 2 \end{bmatrix} * \begin{bmatrix} X \\ Y \end{bmatrix}$$
$$= \begin{bmatrix} 2 & -1 \\ -5 & 3 \end{bmatrix} * \begin{bmatrix} 7 \\ 9 \end{bmatrix}$$

Chapter 11 Matrices

$$= \begin{bmatrix} 2*7 - 1*9 \\ -5*7 + 3*9 \end{bmatrix}$$

$$= \begin{bmatrix} 5 \\ -8 \end{bmatrix}$$

Thus X = 5 and Y = −8 solves the simultaneous equations.

The next program Simultaneous Equations essentially goes through this process to solve a set of N simultaneous equations in N unknowns. In fact the process is to write the set of simultaneous equations into two matrices A(I,J) and B(I). The example given earlier on would be written as follows.

$$\begin{bmatrix} 3 & 1 \\ 5 & 2 \end{bmatrix} \quad \begin{bmatrix} 7 \\ 9 \end{bmatrix}$$

Row operations are then performed until the matrix on the left becomes the identity matrix. When this is achieved the matrix on the right becomes the solution to the set of simultaneous equations.

```
10 REM SIMULTANEOUS EQUATIONS
20 PRINT CHR$(147),CHR$(154) "SIMULTANEO
US EQUATIONS" CHR$(17)
30 PRINT "THIS PROGRAM SOLVES A SET OF N
 SIMULT-"
40 PRINT "ANEOUS EQUATIONS IN N UNKNOWNS
." CHR$(17)
50 PRINT "ENTER THE NUMBER OF EQUATIONS.
" CHR$(17)
60 INPUT "VALUE OF N ";N:PRINT
70 IF N<1 OR N<>INT(N) THEN PRINT "NONSE
NSE!! TRY AGAIN.":GOTO 60
80 PRINT "ENTER THE COEFFICIENTS TERM BY
 TERM." CHR$(17)
90 DIM A(N,N),B(N)
100 REM READ IN MATRIX OF COEFFICIENTS
110 FOR I=1 TO N
120 PRINT CHR$(159);"    ROW";I;CHR$(154)
130 FOR J=1 TO N
140 PRINT "COLUMN";J;:INPUT A(I,J)
150 NEXT
160 INPUT "RIGHT HAND TERM";B(I):PRINT
170 NEXT
180 REM CALCULATING
190 PRINT CHR$(147) "THE MATRIX OF COEFF
```

```
ICIENTS:" CHR$(17)
200 FOR I=1 TO N:FOR J=1 TO N:PRINT A(I,
J);:NEXT:PRINT B(I):NEXT
210 PRINT:PRINT "CALCULATING ";
220 X=0:GOSUB 300
230 REM ENDING
240 PRINT:PRINT:PRINT CHR$(154),"ANOTHER
 GO? Y OR N"
250 GET G$:IF G$<>"Y" AND G$<>"N" THEN 2
50
260 IF G$="Y" THEN RUN
270 PRINT CHR$(147) "BYE FOR NOW":END
300 REM CALCULATING
310 X=X+1:Z=X:PRINT "*";
320 IF A(Z,X)=0 THEN Z=Z+1: IF Z<=N THEN
 320
330 IF Z>N THEN PRINT:PRINT:PRINT "NO SO
LUTION!!":RETURN
340 IF Z<>X THEN R=1/A(Z,X):I=X:K=Z:GOSU
B 500
350 IF A(X,X)<>1 THEN R=1/A(X,X):I=X:GOS
UB 550
360 FOR I=1 TO N
370 IF I=X THEN I=I+1:IF I>N THEN 390
380 IF A(I,X)<>0 THEN R=-A(I,X):K=X:GOSU
B 500
390 NEXT
400 IF X<N THEN 310
410 PRINT:PRINT:PRINT "THE SOLUTION:" CH
R$(17)
420 FOR I=1 TO N
430 IF POS(1)+LEN(STR$(B(I)))>38 THEN PR
INT
440 PRINT B(I);
450 NEXT
460 PRINT:RETURN
500 REM ROW OPERATION ADD R TIMES ROW K
TO ROW I
510 FOR J=1 TO N
520 A(I,J)=A(I,J)+R*A(K,J)
530 NEXT
540 B(I)=B(I)+R*B(K):RETURN
550 REM ROW OPERATION MULTIPLY ROW I BY
R
```

```
560 FOR J=1 TO N
570 A(I,J)=R*A(I,J)
580 NEXT
590 B(I)=R*B(I):RETURN

READY.
```

CHAPTER 12
Codes

There is a growing need for secure transmission of data. This has resulted in much research into cryptography — the art of writing in code or cipher. There are two steps involved in cryptography, first encoding the message or data and then decoding the coded message back to the original.

Substitution codes

The simplest ciphers are the substitution ones in which each letter is substituted for something else, usually a letter from the same alphabet. Here is one such example:

encoding→ ←decoding	encoding→ ←decoding
A — A	B — F
C — K	D — P
E — U	F — Z
G — E	H — J
I — O	J — T
K — Y	L — D
M — I	N — N
O — S	P — X
Q — C	R — H
S — M	T — R
U — W	V — B
W — G	X — L
Y — Q	Z — V

The message MESSAGE would be translated into IUMMAEU.

In this cipher, the substitution for each letter was not chosen at random. If the letters of the alphabet are number 0 to 25 then the substitute for letter numbered NUM is calculated as follows

$$NUM*5 - 26*INT(NUM*5/26)$$

The reverse process is achieved by the formula

$$NUM*21 - 26*INT(NUM*21/26)$$

119

The cipher may be listed with the following short program.

```
10 REM SIMPLE CIPHER
20 PRINT CHR$(147),"CODING" CHR$(17)
30 FOR I=0 TO 25
40 J=I*5-26*INT(I*5/26)
50 PRINT CHR$(65+I) " - " CHR$(65+J),
60 NEXT
70 PRINT:PRINT:PRINT,"DECODING" CHR$(17)
80 FOR I=0 TO 25
100 J=I*21-26*INT(I*21/26)
110 PRINT CHR$(65+I) " - " CHR$(65+J),
120 NEXT
```

Essentially to encode a letter we multiply the number of the letter by 5 and ignore multiples of 26. To decode a letter we need to divide by 5, ignoring multiples of 26. This is the same as multiplying by 21 and ignoring multiples of 26 because

$$5 * 21 = 105$$
$$= 1 + 4 * 26$$

In other words if we ignore multiples of 26 then 21 is the reciprocal of 5. Indeed we say that 21 is the inverse of 5 modulo 26.

One obvious defect of this code is that a message such as PLEASE COME QUICKLY would be encoded as XDUAMU KSIU CWIKYDQ. Spaces are left as spaces. To overcome this objection we should include spaces, full stops, commas, digits and perhaps question marks in our list of letters for substitution.

Since we have a computer at our disposal we should use the ASCII characters. The 59 ASCII characters from 31 to 90 are convenient. These include all the letters that we require and in addition 59 is a prime number which will be useful for our purpose. CHR$(31) is a colour code and will not in fact be used.

Encoding will be done essentially by multiplying by 5. More precisely the single character A$ is encoded to C$ as follows.

$$N = 5 * (ASC(A\$) - 31)$$
$$C\$ + CHR\$(31 + N - 59*INT(N/59))$$

Decoding is achieved by multiplying by 12 (the product of 5 and 12 is 60 which is 1 ignoring multiples of 59). To decode the single character C$ we proceed as follows.

$$M = 12 * (ASC(C\$) - 31)$$
$$A\$ = CHR\$(31 + M - 59*INT(M/59))$$

Chapter 12 Codes

Here is a short program, based on the above cipher, which will encode a message or decode one. There is one added feature, you have to enter one of the numbers 2 to 58 for encoding and decoding. The same number must be used for decoding as for encoding. To encode we essentially multiply by the chosen number, say N. To decode we multiply by the reciprocal of N modulo 59, that is by a number M for which N*M is 1 ignoring multiples of 59.

```
10 REM SUBSTITUTION CODE
20 PRINT CHR$(147),CHR$(154) "SUBSTITUTI
ON CODE" CHR$(17)
30 PRINT "THIS PROGRAM WILL ENCODE AND D
ECODE.          "
40 PRINT "ENTER CODE NUMBER" CHR$(17)
50 INPUT "NUMBER 2 TO 58 ";C:PRINT
60 IF C<>INT(C) OR C<2 OR C>58 THEN PRIN
T "TRY AGAIN.":GOTO 50
70 PRINT "DO YOU WANT TO ENCODE (E) OR D
ECODE (D) A MESSAGE?" CHR$(17)
80 INPUT "E OR D ";A$:PRINT
90 IF A$<>"E" AND A$<>"D" THEN PRINT "WH
ICH?":GOTO 80
100 D=1:I=C:B$="DE":IF A$="E" THEN D=C:B
$="EN":GOTO 130
110 C=C+I:D=D+1:IF C-59*INT(C/59)<>1 THE
N 110
120 REM D IS MULTIPLE REQUIRED FOR ENCOD
ING/DECODING
130 PRINT "TYPE YOUR MESSAGE - UP TO 255
 CHARACTERSLONG." CHR$(17)
140 M$="":N$="":L=0
150 GET G$:IF G$="" THEN 150
160 REM MESSAGE IS ENTERED CHARACTER BY
CHARACTER
170 G=ASC(G$)
180 IF G>31 AND G<91 THEN M$=M$+G$:PRINT
 G$;:L=L+1
190 IF G=20 AND L>0 THEN PRINT G$;:L=L-1
:M$=LEFT$(M$,L)
200 IF G<>13 AND L<255 THEN 150
210 PRINT " "
220 REM ENCODING/DECODING MESSAGE
230 FOR I=1 TO L
240 N=ASC(MID$(M$,I,1))-31:N=D*N
```

121

```
250 N$=N$+CHR$(31+N-59*INT(N/59))
260 NEXT
270 PRINT:PRINT "THE " B$ "CODED MESSAGE
    IS:" CHR$(17):PRINT N$
280 PRINT:PRINT CHR$(154),"ANOTHER GO? Y
    OR N"
290 GET G$:IF G$<>"Y" AND G$<>"N" THEN 2
90
300 IF G$="Y" THEN RUN
310 PRINT CHR$(147) "BYE FOR NOW":END

READY.
```

Despite all our efforts the code is easy (for experts) to break or decrypt. The problem with a substitution code is that each character is invariably represented by some other fixed character. Certain English letters and pairs of letters occur much more frequently than others. For instance in normal English text, the letter E occurs about 13% of the time, the letter T occurs about 9%, the letter P about 2% and Q about 0.2%. Armed with such information it is possible to decrypt a substitution code. The number of different substitution codes possible using the 59 characters (ASCII codes 31 to 90) is about $1.4 * 10^{80}$. Nevertheless substitution codes can be decrypted. This illustrates how deceptive the appearance of large numbers of choices can be.

The next section introduces codes which do not always use the same character for a given character.

Matrix codes

We can use matrices to cipher messages. We illustrate the method with an example. Suppose that we want to encode the message PLEASE COME QUICKLY. First we rearrange the message into two rows as below:

```
P E S    O E Q I K Y
L A E C M  U C L .
```

Next, create a two row matrix from these rows by converting the letters into their ASCII codes less 31.

$$\begin{bmatrix} 49 & 38 & 52 & 1 & 48 & 38 & 50 & 42 & 44 & 58 \\ 45 & 34 & 38 & 36 & 46 & 1 & 54 & 36 & 45 & 15 \end{bmatrix}$$

Now premultiply the matrix with the following matrix.

$$\begin{bmatrix} 2 & -1 \\ -5 & 3 \end{bmatrix}$$

The result is:

$$\begin{bmatrix} 2 & -1 \\ -5 & 3 \end{bmatrix} * \begin{bmatrix} 49 & 38 & 52 & 1 & 48 & 38 & 50 & 42 & 44 & 58 \\ 45 & 34 & 38 & 36 & 46 & 1 & 54 & 36 & 45 & 15 \end{bmatrix}$$

$$= \begin{bmatrix} 53 & 42 & 66 & -34 & 50 & 75 & 46 & 48 & 43 & 101 \\ -110 & -88 & -146 & 103 & -102 & -187 & -88 & -102 & -85 & -245 \end{bmatrix}$$

Now convert the numbers in the matrix so that they are between 0 and 58. This is achieved by adding or subtracting multiples of 59 to the numbers in the following way:

N = N − 59*INT(N/59)

The resulting matrix is shown below.

$$\begin{bmatrix} 53 & 42 & 7 & 25 & 50 & 16 & 46 & 48 & 43 & 42 \\ 8 & 30 & 31 & 44 & 16 & 49 & 30 & 16 & 33 & 50 \end{bmatrix}$$

Finally we add 31 to these numbers and look up the corresponding characters by asking for PRINT CHR$(X). The result is shown below.

```
T  I  &  8  Q  /  M  O  J  I
'  =  >  K  /  P  =  /  @  Q
```

The final coded message is

T'I=&>8KQ//PM=O/J@IQ

Notice that the letter E appears three times in the original message. These three Es have been encoded to I, > and /. Thus this cipher is more subtle than the straight substitution code.

A code is no good unless we can decipher messages. The trick now is to reverse the process. At first sight this may look difficult, but we use some mathematics to help us.

Let's decipher the following message.

I=M?''S?$:8M@>'<M

First we write the coded message in two rows.

```
I  M  "  ?  :  M  >  <
=  ?  S  $  8  @  '  M
```

123

Next, calculate the ASCII codes less 31 to produce a matrix.

$$\begin{bmatrix} 42 & 46 & 3 & 32 & 27 & 46 & 31 & 29 \\ 30 & 32 & 52 & 5 & 25 & 33 & 8 & 46 \end{bmatrix}$$

Now premultiply by the following matrix.

$$\begin{bmatrix} 3 & 1 \\ 5 & 2 \end{bmatrix}$$

Notice that this matrix is not the same as the one used for encoding. However, notice that the decoding matrix is the inverse of the encoding matrix as the following calculation shows.

$$\begin{bmatrix} 3 & 1 \\ 5 & 2 \end{bmatrix} * \begin{bmatrix} 2 & -1 \\ -5 & 3 \end{bmatrix}$$

$$= \begin{bmatrix} 3*2 + 1*-5 & 3*-1 + 1*3 \\ 5*2 + 2*-5 & 5*-1 + 2*3 \end{bmatrix}$$

$$= \begin{bmatrix} 1 & 0 \\ 0 & 1 \end{bmatrix}$$

Let's multiply our decoding matrix by the coded message matrix.

$$\begin{bmatrix} 3 & 1 \\ 5 & 2 \end{bmatrix} * \begin{bmatrix} 42 & 46 & 3 & 32 & 27 & 46 & 31 & 29 \\ 30 & 32 & 52 & 5 & 25 & 33 & 8 & 46 \end{bmatrix}$$

$$= \begin{bmatrix} 156 & 170 & 61 & 101 & 106 & 171 & 101 & 133 \\ 270 & 294 & 119 & 170 & 185 & 296 & 171 & 237 \end{bmatrix}$$

Next, add (or subtract) multiples of 59 to make the numbers between 0 and 58.

$$\begin{bmatrix} 38 & 52 & 2 & 42 & 47 & 53 & 42 & 15 \\ 34 & 58 & 1 & 52 & 8 & 1 & 53 & 1 \end{bmatrix}$$

Now add 31 to these numbers and look up the corresponding characters.

```
E  S  !  I  N  T  I  .
A  Y     S  '     T
```

We thus end up with the message EASY! ISN'T IT.

The next program uses such matrices to code and/or decode messages.

```
10 REM MATRIX CIPHER
20 PRINT CHR$(147).CHR$(154) "    MATRIX
CIPHER" CHR$(17)
30 PRINT "THIS PROGRAM WILL ENCODE AND D
ECODE.      "
40 S=2:FOR I=1 TO S:FOR J=1 TO S:READ A(
I,J):NEXT:NEXT
50 PRINT "DO YOU WANT TO ENCODE (E) OR D
ECODE (D) A MESSAGE?" CHR$(17)
60 INPUT "E OR D ";A$:PRINT
70 IF A$<>"E" AND A$<>"D" THEN PRINT "WH
ICH?":GOTO 60
80 B$="EN"
90 IF A$="D" THEN B$="DE":FOR I=1 TO S:F
OR J=1 TO S:READ A(I,J):NEXT:NEXT
100 PRINT "TYPE YOUR MESSAGE - UP TO 255
 CHARACTERSLONG." CHR$(17)
110 M$="":L=0
120 GET G$:IF G$="" THEN 120
130 G=ASC(G$)
140 IF G>31 AND G<91 THEN M$=M$+G$:PRINT
 G$;:L=L+1
150 IF G=20 AND L>0 THEN PRINT G$;:L=L-1
:M$=LEFT$(M$,L)
160 IF G<>13 AND L<255 THEN 120
170 PRINT " ";M=INT(L/S+0.9):DIM B(S,M),
C(S,M)
180 REM ENCODING/DECODING MESSAGE
190 FOR I=1 TO S:FOR J=1 TO M
200 K=S*J+I-S
210 IF K<=L THEN N=ASC(MID$(M$,K,1))-31:
B(I,J)=N-59*INT(N/59)
220 NEXT:NEXT
230 FOR I=2 TO S:IF B(I,M)<1 THEN B(I,M)
=1
240 NEXT
250 FOR I=1 TO S:FOR J=1 TO M
260 FOR K=1 TO S:C(I,J)=C(I,J)+A(I,K)*B(
K,J):NEXT
270 C(I,J)=C(I,J)-59*INT(C(I,J)/59)
280 NEXT:NEXT
290 PRINT:PRINT "THE " B$ "CODED MESSAGE
 IS:" CHR$(17)
```

```
300 FOR J=1 TO M:FOR I=1 TO S
310 PRINT CHR$(31+C(I,J));
320 NEXT:NEXT:PRINT
330 PRINT:PRINT CHR$(154),"ANOTHER GO? Y
 OR N"
340 GET G$:IF G$<>"Y" AND G$<>"N" THEN 3
40
350 IF G$="Y" THEN RUN
360 PRINT CHR$(147) "BYE FOR NOW":END
400 DATA 2,-1,-5,3
410 DATA 3,1,5,2

READY.
```

For this cipher we used a two by two matrix to encode and the inverse of that matrix to decode. In general we can use any matrix and its inverse as long as both matrices have only integers as their entries. Here is another matrix that you could use for encoding.

$$\begin{bmatrix} 0 & 1 \\ 1 & -2 \end{bmatrix}$$

The corresponding decoding matrix is given next.

$$\begin{bmatrix} 2 & 1 \\ 1 & 0 \end{bmatrix}$$

This method of coding could be made more subtle by using 3 by 3 matrices. The message would need to be written out in three rows and the procedure outlined earlier on followed. Here are two matrices that could be used as encoders and decoders.

$$\begin{bmatrix} 1 & 0 & -1 \\ 2 & 1 & 3 \\ 4 & 2 & 5 \end{bmatrix} \begin{bmatrix} 1 & 2 & -1 \\ -2 & -9 & 5 \\ 0 & 2 & -1 \end{bmatrix}$$

The next program uses these matrices to code and/or decode messages. You could insert your own matrices if you wish, but make sure that they are inverse to each other and that all entries are integral.

```
10 REM MATRIX 3 CIPHER
20 PRINT CHR$(147),CHR$(154) "    MATRIX
 CIPHER" CHR$(17)
```

```
30 PRINT "THIS PROGRAM WILL ENCODE AND D
ECODE.          "
40 S=3:FOR I=1 TO S:FOR J=1 TO S:READ A(
I,J):NEXT:NEXT
50 PRINT "DO YOU WANT TO ENCODE (E) OR D
ECODE (D) A MESSAGE?" CHR$(17)
60 INPUT "E OR D ";A$:PRINT
70 IF A$<>"E" AND A$<>"D" THEN PRINT "WH
ICH?":GOTO 60
80 B$="EN"
90 IF A$="D" THEN B$="DE":FOR I=1 TO S:F
OR J=1 TO S:READ A(I,J):NEXT:NEXT
100 PRINT "TYPE YOUR MESSAGE - UP TO 255
 CHARACTERSLONG." CHR$(17)
110 M$="":L=0
120 GET G$:IF G$="" THEN 120
130 G=ASC(G$)
140 IF G>31 AND G<91 THEN M$=M$+G$:PRINT
 G$;:L=L+1
150 IF G=20 AND L>0 THEN PRINT G$;:L=L-1
:M$=LEFT$(M$,L)
160 IF G<>13 AND L<255 THEN 120
170 PRINT " ":M=INT(L/S+0.9):DIM B(S,M),
C(S,M)
180 REM ENCODING/DECODING MESSAGE
190 FOR I=1 TO S:FOR J=1 TO M
200 K=S*J+I-S
210 IF K<=L THEN N=ASC(MID$(M$,K,1))-31:
B(I,J)=N-59*INT(N/59)
220 NEXT:NEXT
230 FOR I=2 TO S:IF B(I,M)<1 THEN B(I,M)
=1
240 NEXT
250 FOR I=1 TO S:FOR J=1 TO M
260 FOR K=1 TO S:C(I,J)=C(I,J)+A(I,K)*B(
K,J):NEXT
270 C(I,J)=C(I,J)-59*INT(C(I,J)/59)
280 NEXT:NEXT
290 PRINT:PRINT "THE " B$ "CODED MESSAGE
 IS:" CHR$(17)
300 FOR J=1 TO M:FOR I=1 TO S
310 PRINT CHR$(31+C(I,J));
320 NEXT:NEXT:PRINT
330 PRINT:PRINT CHR$(154),"ANOTHER GO? Y
```

```
 OR N"
340 GET G$:IF G$<>"Y" AND G$<>"N" THEN 3
40
350 IF G$="Y" THEN RUN
360 PRINT CHR$(147) "BYE FOR NOW":END
400 DATA 1,0,-1,2,1,3,4,2,5
410 DATA 1,2,-1,-2,-9,5,0,2,-1

READY.
```

Public-key codes

The codes described in the previous section have a defect. Once you know how to encode a message you also know how to decode it. Public-key cryptosystems are different. They come in two parts: the encoding key, which is made public, enabling anyone to encode messages; and the decoding key, which is kept secret, enabling only the originator of the code to decode messages.

We now describe a public-key system. First find two very large prime numbers P and Q, each should have about 50 decimal digits making this proposition impractical for your Commodore 64. Let N be the product of P and Q. Now choose an integer A which is less than N and has no factor common with $(P-1)*(Q-1)$. You may now publicly announce the numbers N and A.

How is a message encoded with the numbers N and A? This is performed as follows.

1. Translate the message into numbers (space = 01, A = 34, etc.); the message is then one large number.
2. Take the message in number form and break it up into blocks of a convenient size.
3. Encode each block B as follows:

   $C = B{\uparrow}A - N*INT((B{\uparrow}A)/N)$

the number B is raised to the power A and multiples of N are removed to make the resulting number between 0 and N.

How is the resulting message decoded? Since the greatest common divisor of A and $(P-1)*(Q-1)$ is 1 we can find two numbers X and Y so that $A*X + (P-1)*(Q-1)*Y = 1$. Using X we can decode the message as follows.

1. Break the coded message into blocks.
2. For each block C perform the following calculation:

   $C{\uparrow}X - N*INT((C{\uparrow}X)/N).$

Chapter 12 Codes

3. Join the resulting blocks back again and decode numbers back to characters.

The process works because

$$(B \uparrow A) \uparrow X = B \uparrow (A*X)$$
$$= B \uparrow (1 - (P-1)*(Q-1)*Y)$$
$$= B + \text{multiples of } N.$$

The last statement follows from a theorem proved by the mathematician Fermat.

Why is the code difficult to break? Notice that to decode the message we need to know the value of X. This can be calculated from the value of $(P-1)*(Q-1)$. In order to know $(P-1)*(Q-1)$ we need to know P and Q. Publicly only A and N have been announced. In theory, once we know N we can factorise it to find P and Q. However N is about 100 decimal digits long and factorisation of such numbers takes an enormous amount of computer time to perform (several millions of years). It is on this basis that the cipher is safe.

All that remains for you to produce a very secure code is to write a program for your Commodore 64 which is capable of handling very long numbers precisely. See the chapter entitled Odds and Ends.

CHAPTER 13
Random!

Heads and tails

When a coin is tossed it is equally likely that the 'head' or 'tail' shows, at least if the coin is a fair coin. Your Commodore 64 can simulate the tossing of a coin by using the inbuilt RND function.

For any positive number X, RND(X) returns a (pseudo) random number in the range from 0 up to 1 (but not including 1). If you switch on your Commodore 64 and type the following:

FOR I = 1 TO 5:PRINT RND(1):NEXT I

you will get the following sequence of numbers:

.185564016
.0468986348
.827743801
.554749226
.897233831

If you switch your 64 off and then on again (not recommended) and repeat the instruction

FOR I = 1 TO 5:PRINT RND(1):NEXT I

you will get the same sequence of numbers. To overcome this it is usual to start any program involving random numbers with a line like

Y = RND(−TI)

which has the effect of starting a new sequence of random numbers. In general, if X is negative then RND(X) starts a new sequence of random numbers. If X is 0 then the resulting number is the same as the last one.

The following short program simulates the tossing of a fair coin. A list is printed out showing whether a head (H) or tail (T) appears. 100 such letters are printed and a count of the number of heads and tails is displayed.

```
10 REM HEADS AND TAILS
20 PRINT CHR$(147),"HEADS AND TAILS" CHR
```

131

```
$(17)
30 PRINT "THIS PROGRAM SIMULATES THE TOS
SING OF A FAIR COIN 100 TIMES." CHR$(17)
40 PRINT,"     PRESS Y":K=0
50 GET G$:IF G$<>"Y" THEN 50
60 Y=RND(-TI)
70 K=K+1:J=0:PRINT,CHR$(145) CHR$(158) "
 RUN NUMBER" STR$(K) "      " CHR$(17)
80 FOR I=1 TO 100
90 A$="H":IF RND(1)>=0.5 THEN A$="T":J=J
+1
100 PRINT A$;
110 NEXT
120 PRINT:PRINT:PRINT "NUMBER OF HEADS";
100-J;"TAILS";J
130 PRINT:PRINT CHR$(154),"ANOTHER GO? Y
 OR N"
140 GET G$:IF G$<>"Y" AND G$<>"N" THEN 1
40
150 IF G$="Y" THEN 60
160 PRINT CHR$(147) "BYE FOR NOW":END

READY.
```

Change line 60 to the following line.

60 Y = RND(−1)

You should now notice that the same sequence of heads and tails appears every time the program is run. That's why we use

60 Y = RND(−TI)

to randomise the sequence of random numbers.

Of dice and men

If a fair six-sided die is thrown then one of the six numbers 1, 2, 3, 4, 5 and 6 will appear, none more likely than another. Once again, the Commodore 64 may be used to simulate dice throwing. The next program illustrates the result of rolling a die one hundred and twenty times. A count of each number thrown is printed at the end.

```
10 REM DIE ROLLING
20 PRINT CHR$(147)."   DIE ROLLING" CHR$(
```

```
17)
30 PRINT "THIS PROGRAM SIMULATES THE ROL
LING OF A FAIR DIE 120 TIMES." CHR$(17)
40 PRINT,"     PRESS Y":K=0
50 GET G$:IF G$<>"Y" THEN 50
60 Y=RND(-TI)
70 K=K+1:FOR I=1 TO 6:A(I)=0:NEXT
80 PRINT,CHR$(145) CHR$(158) " RUN NUMBE
R" STR$(K) "        " CHR$(17)
90 FOR I=1 TO 100
100 L=INT(RND(1)*6)+1:PRINT STR$(L);:A(L
)=A(L)+1
110 NEXT
120 PRINT:PRINT:FOR I=1 TO 6:PRINT "NUMB
ER OF";STR$(I);"'S =";A(I):NEXT
130 PRINT:PRINT CHR$(154),"ANOTHER GO? Y
 OR N"
140 GET G$:IF G$<>"Y" AND G$<>"N" THEN 1
40
150 IF G$="Y" THEN 60
160 PRINT CHR$(147) "BYE FOR NOW":END

READY.
```

Another version of the program appears next. This one includes a display of the face of the die. The design for the faces is contained in the array D$(I,J).

```
10 REM PICTURE DIE
20 PRINT CHR$(147),"DIE ROLLING WITH DIS
PLAY" CHR$(17)
30 PRINT "THIS PROGRAM SIMULATES THE ROL
LING OF A FAIR DIE." CHR$(17)
40 FOR I=0 TO 6:FOR J=0 TO 6:READ A:A$(I
)=A$(I)+CHR$(A):NEXT:NEXT
50 FOR I=1 TO 6:FOR J=0 TO 6:READ A:D$(I
,J)=A$(A):NEXT:NEXT
60 PRINT,"     PRESS Y":K=0
70 GET G$:IF G$<>"Y" THEN 70
80 Y=RND(-TI)
90 K=K+1:PRINT,CHR$(145) CHR$(158) "   R
OLL NUMBER" STR$(K) "     " CHR$(17)
100 L=INT(RND(1)*6)+1:A(L)=A(L)+1
110 FOR J=0 TO 7:PRINT TAB(16) D$(L,J):N
```

```
EXT
120 PRINT:PRINT:PRINT "COUNT";:FOR I=1 T
O 3:PRINT STR$(I);"'S =";A(I);:NEXT
130 PRINT:PRINT "        ";:FOR I=4 TO 6:PR
INT STR$(I);"'S =";A(I);:NEXT:PRINT
140 PRINT:PRINT:PRINT:PRINT CHR$(154),"A
NOTHER GO? Y OR N"
150 GET G$:IF G$<>"Y" AND G$<>"N" THEN 1
50
160 IF G$="Y" THEN 80
170 PRINT CHR$(147) "BYE FOR NOW":END
200 REM DATA FOR DESIGN
210 DATA 207,183,183,183,183,183,208
220 DATA 180,32,32,32,32,32,170
230 DATA 180,113,32,32,32,32,170
240 DATA 180,32,32,113,32,32,170
250 DATA 180,32,32,32,32,113,170
260 DATA 180,113,32,32,32,113,170
270 DATA 204,175,175,175,175,175,186
300 REM DATA FOR EACH DIE FACE
310 DATA 0,1,1,3,1,1,6
320 DATA 0,4,1,1,1,2,6
330 DATA 0,4,1,3,1,2,6
340 DATA 0,5,1,1,1,5,6
350 DATA 0,5,1,3,1,5,6
360 DATA 0,5,1,5,1,5,6

READY.
```

Rolling two dice simultaneously can be simulated just as easily. The possible score on each roll is one of the numbers from 2 to 12. As you are no doubt aware, some scores are more likely to occur than others. This fact should become apparent with either of the next two programs. If two dice are rolled several times then the expected proportion of time (or probability) that each score occurs is given in the next table.

Score	Probability
2	1/36
3	2/36
4	3/36
5	4/36
6	5/36
7	6/36
8	5/36
9	4/36

Chapter 13 Random

10	3/36
11	2/36
12	1/36

```
10 REM TWO DICE
20 PRINT CHR$(147)," TWO DICE ROLLING" CHR$(17)
30 PRINT "THIS PROGRAM SIMULATES THE ROLLING OF  TWO FAIR DICE." CHR$(17)
40 PRINT,"    PRESS Y":K=0
50 GET G$:IF G$<>"Y" THEN 50
60 Y=RND(-TI)
70 K=K+1:PRINT,CHR$(145) CHR$(158) " RUN NUMBER" STR$(K) "    " CHR$(17)
80 FOR I=1 TO 60
90 L=INT(RND(1)*6)+1:M=INT(RND(1)*6)+1:A(L+M-2)=A(L+M-2)+1
100 PRINT CHR$(154);"**";CHR$(158);L;M;
110 NEXT
120 PRINT:PRINT "SCORE COUNT"
130 FOR I=0 TO 10
140 PRINT STR$(I+2);"'S =";A(I);:IF I=2 OR I=5 OR I=8 THEN PRINT
150 NEXT
160 PRINT:PRINT:PRINT:PRINT CHR$(154),"CONTINUE?  Y OR N"
170 GET G$:IF G$<>"Y" AND G$<>"N" THEN 170
180 IF G$="Y" THEN 60
190 PRINT CHR$(147) "BYE FOR NOW":END

READY.

10 REM PICTURE TWO DICE
20 PRINT CHR$(147) "    TWO DICE ROLLING WITH DISPLAY" CHR$(17)
30 PRINT "THIS PROGRAM SIMULATES THE ROLLING OF  TWO FAIR DICE." CHR$(17)
40 FOR I=0 TO 6:FOR J=0 TO 6:READ A:A$(I)=A$(I)+CHR$(A):NEXT:NEXT
50 FOR I=1 TO 6:FOR J=0 TO 6:READ A:D$(I,J)=A$(A):NEXT:NEXT
60 PRINT,"    PRESS Y":K=0
70 GET G$:IF G$<>"Y" THEN 70
```

```
80 Y=RND(-TI)
90 K=K+1:PRINT,CHR$(145) CHR$(158) "     R
OLL NUMBER" STR$(K) "      " CHR$(17)
100 L=INT(RND(1)*6)+1:M=INT(RND(1)*6)+1:
A(L+M-2)=A(L+M-2)+1
110 FOR J=0 TO 7:PRINT TAB(10) D$(L,J) S
PC(6) D$(M,J):NEXT
120 PRINT:PRINT "SCORE COUNT"
130 FOR I=0 TO 10
140 PRINT STR$(I+2);"'S =";A(I);:IF I=3
OR I=7 THEN PRINT
150 NEXT
160 PRINT:PRINT:PRINT:PRINT CHR$(154),"A
NOTHER GO? Y OR N"
170 GET G$:IF G$<>"Y" AND G$<>"N" THEN 1
70
180 IF G$="Y" THEN 80
190 PRINT CHR$(147) "BYE FOR NOW":END
200 REM DATA FOR DESIGN
210 DATA 207,183,183,183,183,183,208
220 DATA 180,32,32,32,32,32,170
230 DATA 180,113,32,32,32,32,170
240 DATA 180,32,32,113,32,32,170
250 DATA 180,32,32,32,32,113,170
260 DATA 180,113,32,32,32,113,170
270 DATA 204,175,175,175,175,175,186
300 REM DATA FOR EACH DIE FACE
310 DATA 0,1,1,3,1,1,6
320 DATA 0,4,1,1,1,2,6
330 DATA 0,4,1,3,1,2,6
340 DATA 0,5,1,1,1,5,6
350 DATA 0,5,1,3,1,5,6
360 DATA 0,5,1,5,1,5,6

READY.
```

Playing cards

A regular pack of playing cards has 52 cards. In a well shuffled pack any one of the 52 cards is likely to appear at the top. The next program, Cards, illustrates how the Commodore 64 can simulate picking a card from a well shuffled pack of playing cards. On each new selection it is assumed that the previously selected card is replaced and the cards are well shuffled again.

Chapter 13 Random

```
10 REM CARDS
20 POKE 53281,15:PRINT CHR$(147),CHR$(31
)  "       CARDS" CHR$(17)
30 PRINT "THIS PROGRAM SIMULATES THE DRA
WING OF"
40 PRINT "A CARD FROM A WELL SHUFFLED PA
CK." CHR$(17)
50 DIM A$(12):FOR I=0 TO 12:READ A$(I):N
EXT
60 FOR I=0 TO 3:READ A:B$(I)=CHR$(A):NEX
T
70 FOR I=0 TO 3:READ A:C$(I)=CHR$(A):NEX
T
80 PRINT,"    PRESS Y":K=0
90 GET G$:IF G$<>"Y" THEN 90
100 Y=RND(-TI)
110 K=K+1:PRINT,CHR$(145) "SELECTION NUM
BER" STR$(K) "      " CHR$(17)
120 L=INT(RND(1)*13):M=INT(RND(1)*4)
130 PRINT TAB(16+(L=9)) C$(M) A$(L) "  "
B$(M)
140 PRINT:PRINT CHR$(31),"ANOTHER GO? Y
OR N"
150 GET G$:IF G$<>"Y" AND G$<>"N" THEN 1
50
160 IF G$="Y" THEN 100
170 PRINT CHR$(147) "BYE FOR NOW":END
200 REM DATA
210 DATA A,2,3,4,5,6,7,8,9,10,J,Q,K
220 DATA 120,122,115,97
230 DATA 144,28,28,144
READY.
```

In the program Cards the computer first selects a number between 0 and 12 which (by adding 1) determines the number appearing on the card. Next, it selects a number between 0 and 3, this determines which of the four suits the card represents. An alternative way would be to select a number between 0 and 51:

$K = INT(RND(1)*52)$

and pick off the suit and number of the card from this number. This is achieved by the following line.

$L = INT(K/4) : M = K - 4*L$

The number L now determines the card number while M determines the suit.

The program Cards selects one card from a well shuffled pack. With each new selection the card is replaced and the pack reshuffled. What if we want to shuffle the pack once and then list the cards as they appear in sequence from top to bottom? A different routine is required to achieve this. Essentially we number the cards 0 to 51 and then randomly rearrange these 52 numbers. This rearranging is done systematically; first the first number is exchanged randomly with one of the other 51 numbers. Then the second number is exchanged with one of the remaining 50 numbers. And so on. The program Card Shuffle shows the technique.

```
10 REM CARD SHUFFLE
20 POKE 53281,15:PRINT CHR$(147),CHR$(31
)  "     CARD SHUFFLE" CHR$(17)
30 PRINT "THIS PROGRAM ILLUSTRATES THE S
HUFFLING"
40 PRINT "OF A PACK OF CARDS." CHR$(17)
50 DIM A$(12):FOR I=0 TO 12:READ A$(I):N
EXT
60 FOR I=0 TO 3:READ A:B$(I)=CHR$(A):NEX
T
70 FOR I=0 TO 3:READ A:C$(I)=CHR$(A):NEX
T
80 DIM D%(51):FOR I=0 TO 51:D%(I)=I:NEXT
90 PRINT,"       PRESS Y" CHR$(145):K=0
100 GET G$:IF G$<>"Y" THEN 100
110 Y=RND(-TI)
120 K=K+1:PRINT,"  SHUFFLE NUMBER" STR$(K
)  "   " CHR$(17)
130 REM MIXING
140 FOR I=0 TO 50
150 L=INT(RND(1)*(52-I))+I
160 REM L SATISFIES I <= L <= N
170 T=D%(I):D%(I)=D%(L):D%(L)=T
180 NEXT
190 FOR I=0 TO 51
200 L=INT(D%(I)/4):M=D%(I)-4*L:IF L<>9 T
HEN PRINT " ";
210 PRINT C$(M) A$(L) "   " B$(M),;
220 NEXT
230 PRINT:PRINT:PRINT CHR$(31),"ANOTHER
GO? Y OR N":PRINT:PRINT:PRINT
240 GET G$:IF G$<>"Y" AND G$<>"N" THEN 2
```

Chapter 13 Random

```
40
250 IF G$="Y" THEN 120
260 PRINT CHR$(147) "BYE FOR NOW":END
270 REM DATA
300 DATA A,2,3,4,5,6,7,8,9,10,J,Q,K
310 DATA 120,122,115,97
320 DATA 144,28,28,144

READY.
```

Non-equally likely events

Most of the examples we have looked at so far have the property that any one of the events that can occur is as likely as any one of the others. The next example is not.

A bucket contains 100 coloured buttons. There are 6 red buttons, 54 blue ones and 40 green ones. To simulate the selection of a button from the bucket we use the following lines.

>X = RND(1)
>R$ = "RED"
>IF X > = 0.06 THEN R$ = "BLUE"
>IF X > = 0.60 THEN R$ = "GREEN"
>PRINT R$

The program Buttons simulates the selection of a button from the bucket. 100 selections are made, after each selection the button is replaced.

```
10 REM BUTTONS
20 POKE 53281,15:PRINT CHR$(147),CHR$(15
4) "       BUTTONS" CHR$(17)
30 PRINT "THIS PROGRAM ILLUSTRATES PICKI
NG A"
40 PRINT "BUTTON FROM A BUCKET WITH 6 RE
D, 54 BLUE";
50 PRINT "AND 40 GREEN BUTTONS." CHR$(17
)
60 PRINT,"      PRESS Y":K=0
70 GET G$:IF G$<>"Y" THEN 70
80 Y=RND(-TI)
90 K=K+1:PRINT,CHR$(145) "    RUN NUMBER"
   STR$(K) "   " CHR$(17)
100 X=RND(1):R$=CHR$(28)+CHR$(166)+"  RE
D"
110 IF X>=0.06 THEN R$=CHR$(31)+CHR$(166
```

139

```
)+"    BLUE"
120 IF X>=0.60 THEN R$=CHR$(30)+CHR$(166
)+"    GREEN"
130 PRINT TAB(16) R$:PRINT:PRINT CHR$(15
4) "COUNT:",:
140 S=ASC(R$)-28:A(S)=A(S)+1
150 PRINT CHR$(28) "RED";A(0),:
160 PRINT CHR$(31) "BLUE";A(3),:
170 PRINT CHR$(30) "GREEN";A(2)
180 PRINT:PRINT CHR$(154),"ANOTHER GO? Y
 OR N"
190 GET G$:IF G$<>"Y" AND G$<>"N" THEN 1
90
200 IF G$="Y" THEN 90
210 PRINT CHR$(147) "BYE FOR NOW":POKE 5
3281,6:END
```

READY.

CHAPTER 14
Meaningful data

What do you do when presented with a large amount of numerical data? For instance, here's a set of numbers which could be the results of some experiment, results of some examination, etc.

```
23  67  89  45  40  10   5
19  99  40  23   9  11  21
34  34  56  41  42  27  80
```

You can gain a great deal of information about the data by looking at such descriptive measures as the mean, the variance and the standard deviation. But first the data has to be entered into our computer.

Such data could be stored on your Commodore 64 by using an array X(I), so that X(0) = 23, X(1) = 67, etc. It could be entered interactively by a simple program. This is illustrated with the programs Data Entry I and Data Entry II.

```
200 REM DATA ENTRY I
210 PRINT CHR$(147).CHR$(154) "        DATA ENTRY" CHR$(17)
220 PRINT "THIS ALLOWS YOU TO ENTER SOME NUMERICAL DATA (AT LEAST 2)." CHR$(17)
230 PRINT "HOW MUCH DATA DO YOU WANT TO ENTER?" CHR$(17)
240 INPUT "NUMBER ";N:PRINT
250 IF N<2 OR N<>INT(N) THEN PRINT "BE R EASONABLE!":GOTO 240
260 N=N-1:DIM X(N)
270 PRINT "NOW ENTER THE" N+1 "ITEMS OF DATA." CHR$(17)
280 FOR I=0 TO N
290 PRINT "DATA NUMBER";I+1;:INPUT X(I)
300 NEXT
310 PRINT:PRINT."PRESS Y TO CONTINUE"
320 GET G$:IF G$<>"Y" THEN 320

READY.
```

141

Data Entry I asks, at the start, for the amount of data to be entered. The array X(I) is then dimensioned appropriately. On the other hand Data Entry II assumes that there are no more than 100 numbers to be entered. You enter the data when requested, to stop you enter -99999. If necessary, the number M in line 240 may be changed from 100 to some other number.

This chapter contains several short programs which may be added to the Data Entry programs. The final result is a useful program which will help analyse your data.

```
200 REM DATA ENTRY II
210 PRINT CHR$(147),CHR$(154) "        DATA
ENTRY" CHR$(17)
220 PRINT "THIS ALLOWS YOU TO ENTER SOME
 NUMERICAL DATA (AT LEAST 2)." CHR$(17)
230 PRINT "ENTER YOUR DATA, ITEM BY ITEM
." CHR$(17)
240 M=100:DIM X(M)
250 FOR I=0 TO M
260 PRINT::IF I>1 THEN PRINT,CHR$(154) S
PC((I+1)/10) " TYPE -99999 TO END."
270 PRINT CHR$(158) "DATA NUMBER";I+1;
280 INPUT X(I):IF X(I)=-99999 AND I>1 TH
EN N=I-1:I=M
290 IF X(I)=-99999 AND I<2 THEN PRINT CH
R$(154) "TOO EARLY TO END.":I=I-1
300 NEXT
310 PRINT:PRINT,CHR$(154) "PRESS Y TO CO
NTINUE."
320 GET G$:IF G$<>"Y" THEN 320
```

The mean

The *mean* or average is an important statistical measure. It is obtained by adding all the numbers together and dividing the sum by the number of numbers.

$$\text{mean} = \frac{\text{sum of data}}{\text{number of data}}$$

If the numbers are stored in the array X(I) for I = 0 to N $-$ 1 then the mean XM may be calculated using the following program lines.

Chapter 14 Meaningful Data

```
X = 0
FOR I = 0 TO N − 1 : X = X + X(I) : NEXT
XM = X/N
```

The program Mean is a short program that calculates the mean and prints it out. You may incorporate it with one of the Data Entry programs.

```
400 REM MEAN OF DATA
410 PRINT CHR$(147),CHR$(154) "   DATA AN
ALYSIS" CHR$(17) CHR$(158)
415 PRINT "NUMBER OF DATA ITEMS = ";N+1;
CHR$(17)
420 X=0:FOR I=0 TO N:X=X+X(I):NEXT:XM=X/
(N+1)
430 PRINT "MEAN = ";XM;CHR$(17)
```

Max, min and spread

It's often useful to know the maximum and minimum values of data. A simple search may be made by your Commodore 64 to find these. The following program is a short program which performs this function. In addition, the *range* or *spread* of the data is calculated. This is simply the difference between the largest and smallest numbers in the data.

```
500 REM MAX, MIN AND SPREAD OF DATA
510 MAX=-10 E 37:MIN=10 E 37
520 FOR I=0 TO N
530 IF X(I)>MAX THEN MAX=X(I)
540 IF X(I)<MIN THEN MIN=X(I)
550 NEXT
560 PRINT "MINIMUM VALUE = ";MIN
570 PRINT "MAXIMUM VALUE = ";MAX
580 PRINT "THE SPREAD IS = ";MAX-MIN;CHR
$(17)
```

You may want your data sorted into increasing (or decreasing) order. Several methods (such as bubble sort, quick sort, shell sort etc.) are available. Details are not given here.

Standard deviation and variance

The mean is a simple, useful and powerful tool. But it does not tell us all we need to know. For instance, look at the following sets of data:

DATA for X(I)   20, 21, 20, 19
DATA for Y(I)   38, 26, 14, 2

The values of the means XM and YM are both 20. However there is much greater variation in the data for Y(I) than for X(I). This variation may be measured by using the standard deviation of the data.

The *standard deviation* of a set of data is given by the following formula.

$$\frac{\text{standard}}{\text{deviation}} = \text{SQR}\left[\frac{\text{sum (difference between data and mean)}^2}{\text{number of data} - 1}\right]$$

The *variance* is the square of the standard deviation.

The procedure for calculating the standard deviation of the data stored in the array X(I) is as follows.

1. Calculate the mean XM.
2. Find the deviations from the mean, that is, the values $X(I) - XM$.
3. Square each deviation, that is, find the values $(X(I) - XM)^2$.
4. Sum the squares of the deviations.
5. Divide by the number of terms less 1. This gives the variance XV of the data.
6. Take the square root. This is the standard deviation XD of the data.

The standard deviation provides an idea of how much the data is dispersed or spread out around the mean. Look at the following examples again.

DATA for X(I)   20, 21, 20, 19
DATA for Y(I)   38, 26, 14, 2

The standard deviations are given by the following calculations.

$$\begin{aligned}
XD &= \text{SQR}((0*0 + 1*1 + 0*0 + (-1)*(-1))/3) \\
&= \text{SQR}(2/3) \\
&\quad 0.816496581 \\
YD &= \text{SQR}((18*18 + 6*6 + (-6)*(-6) + (-18)*(-18))/3) \\
&= \text{SQR}(240) \\
&= 15.4919334
\end{aligned}$$

This certainly reflects the difference in the spread of the data.

The next short program is for calculating the standard deviation of the data stored in an array X(I).

```
600 REM STANDARD DEVIATION
610 X=0:FOR I=0 TO N:Y=X(I)-XM:X=X+Y*Y:NEXT
620 XD=SQR(X/N)
630 PRINT "STANDARD DEVIATION = ";XD;CHR$(17)
```

Chapter 14 Meaningful Data

Confidence intervals

The standard deviation is useful because it indicates to what extent the data is spread about the mean. In many mass manufacturing processes the product produced varies slightly in size or quality or length etc. We refer to the items we are measuring as the population. The variation in the population is often normally distributed.

Statisticians have found that the normal curve or normal distribution approximates a large number of real-life data. If the amount of data is large (more than about 30) then it is often assumed, for calculations, that the population is normally distributed, even though it may not be normally distributed.

Roughly speaking a population is normally distributed if it is symmetrically spread about the mean; with most of the population at the mean and very little far away from the mean.

More precisely, in a normally distributed population about 68% of the population lies within 1 standard deviation from the mean, and about 96% lies within 2 standard deviations from the mean. More generally, the next table lists the percentages associated with various multiples of the standard deviation.

% of population	Multiple of standard deviation
50%	0.6745
68.27%	1
80%	1.28
90%	1.645
95%	1.96
95.45%	2
99%	2.575
99.73%	3

The above table shows that we would expect 95% of a population to be within 1.96 times the standard deviation from the mean. In other words 95% of the population lies within the range

XM − 1.96*XD to XM + 1.96*XD

where XM is the mean and XD the standard deviation. We refer to this interval as the 95% confidence interval for the population. Similarly, the 99% confidence interval for the population is from

XM − 2.575*XD to XM + 2.575*XD

145

Here is an illustration of how confidence intervals may be useful. You suspect that a grocer is selling incompletely filled 2 litre bottles of lemonade. You buy ten bottles and measure their contents carefully. The results in litres are:

2.001, 2.040, 2.020, 2.000, 2.015
2.006, 2.005, 2.031, 2.008, 2.018

All the bottles contain 2 litres or more. But, let's calculate the mean and standard deviation of this data. The results are:

XM = 2.0144
XD = 0.0132

Now, assuming that our sample came from a normally distributed population we can calculate some confidence intervals. The 95% confidence interval for the population is from

2.0144 − 1.96*0.0132 to 2.0144 + 1.96*0.0132

that is, from

1.989 to 2.040

Thus we expect 95% of the bottles to contain between 1.989 and 2.040 litres. This means that we expect 2.5% would contain more than 2.040 litres, while 2.5% would contain less than 1.989 litres. We could also work out the 90% confidence interval. This is from

2.0144 − 1.645*0.0132 to 2.0144 − 1.645*0.0132

or, from

1.993 to 2.036

Thus 90% of the bottles are expected to have between 1.993 and 2.036 litres. It follows that at least 5% of the lemonade bottles would contain less than the required 2 litres. (Equally, at least 5% contain more than 2.036 litres.)

In the example just given many assumptions have been made and the conclusions reached would be insufficient to lead to legal proceedings.

The mean of our population was calculated from a sample. How do we know that this is the actual mean of the population? The mean may vary with the sample taken. But, we can estimate how far our sample mean is from the real mean by using the standard deviation. We can say with 95% confidence, that the mean is from

XM − 1.96*XD/SQR(N−1) to XM + 1.96*XD/SQR(N−1)

Chapter 14 Meaningful Data

where XM is the mean calculated from the sample of size N. We call this the 95% confidence interval for the mean. The 99% confidence interval for the mean is given by

XM − 2.575*XD/SQR(N−1) to XM + 2.575*XD/SQR)(N−1)

Strictly speaking these calculations are valid if the sample size is large (say greater than 30). For smaller samples we should be using what is called the *student's t distribution* instead of the normal distribution. But this is beyond the scope of this book.

Don't take the confidence intervals too seriously and don't confuse the two types of confidence intervals.

The next program calculates 95% and 99% confidence intervals for the population and mean.

```
700 REM CONFIDENCE INTERVALS
710 C1=1.96*XD:M1=C1/SQR(N):C2=2.575*XD:
M2=C2/SQR(N)
720 PRINT "95% CONFIDENCE INTERVALS:"
730 PRINT "POP. FROM ";XM-C1;" TO ";XM+C
1
740 PRINT "MEAN FROM ";XM-M1;" TO ";XM+M
1;CHR$(17)
750 PRINT "99% CONFIDENCE INTERVALS:"
760 PRINT "POP. FROM ";XM-C2;" TO ";XM+C
2
770 PRINT "MEAN FROM ";XM-M2;" TO ";XM+M
2;CHR$(17)
```

Finishing touches

The programs in this chapter may be linked together to produce a useful program for analysing data. The following program provides the necessary linking parts.

```
10 REM DATA ANALYSIS
20 PRINT CHR$(147),CHR$(154) "   DATA AN
ALYSIS" CHR$(17) CHR$(158)
30 PRINT "THIS PROGRAM ALLOWS YOU TO ENT
ER DATA   AND ANALYSE IT." CHR$(17)
40 PRINT "AFTER THE DATA HAS BEEN ENTERE
D YOU WILL";
```

147

```
50 PRINT "BE PROVIDED WITH THE FOLLOWING
 INFORMA- TION." CHR$(17)
60 PRINT "1. MEAN OF DATA" CHR$(17)
70 PRINT "2. MAX, MIN AND SPREAD OF DATA
" CHR$(17)
80 PRINT "3. STANDARD DEVIATION" CHR$(17
)
90 PRINT "4. 95% AND 99% CONFIDENCE INTE
RVALS" CHR$(17)
100 PRINT:PRINT,CHR$(154) "PRESS Y TO CO
NTINUE."
110 GET G$:IF G$<>"Y" THEN 110
200 REM DATA ENTRY II
210 PRINT CHR$(147),CHR$(154) "    DATA
ENTRY" CHR$(17)
220 PRINT "THIS ALLOWS YOU TO ENTER SOME
 NUMERICAL DATA (AT LEAST 2)." CHR$(17)
230 PRINT "ENTER YOUR DATA, ITEM BY ITEM
." CHR$(17)
240 M=100:DIM X(M)
250 FOR I=0 TO M
260 PRINT::IF I>1 THEN PRINT,CHR$(154) S
PC((I+1)/10) " TYPE -99999 TO END."
270 PRINT CHR$(158) "DATA NUMBER";I+1;
280 INPUT X(I):IF X(I)=-99999 AND I>1 TH
EN N=I-1:I=M
290 IF X(I)=-99999 AND I<2 THEN PRINT CH
R$(154) "TOO EARLY TO END.":I=I-1
300 NEXT
310 PRINT:PRINT,CHR$(154) "PRESS Y TO CO
NTINUE."
320 GET G$:IF G$<>"Y" THEN 320
400 REM MEAN OF DATA
410 PRINT CHR$(147),CHR$(154) "    DATA AN
ALYSIS" CHR$(17) CHR$(158)
415 PRINT "NUMBER OF DATA ITEMS = ";N+1;
CHR$(17)
420 X=0:FOR I=0 TO N:X=X+X(I):NEXT:XM=X/
(N+1)
430 PRINT "MEAN = ";XM;CHR$(17)
500 REM MAX, MIN AND SPREAD OF DATA
510 MAX=-10 E 37:MIN=10 E 37
520 FOR I=0 TO N
530 IF X(I)>MAX THEN MAX=X(I)
```

```
540 IF X(I)<MIN THEN MIN=X(I)
550 NEXT
560 PRINT "MINIMUM VALUE = ";MIN
570 PRINT "MAXIMUM VALUE = ";MAX
580 PRINT "THE SPREAD IS = ";MAX-MIN:CHR$(17)
600 REM STANDARD DEVIATION
610 X=0:FOR I=0 TO N:Y=X(I)-XM:X=X+Y*Y:NEXT
620 XD=SQR(X/N)
630 PRINT "STANDARD DEVIATION = ";XD:CHR$(17)
700 REM CONFIDENCE INTERVALS
710 C1=1.96*XD:M1=C1/SQR(N):C2=2.575*XD:M2=C2/SQR(N)
720 PRINT "95% CONFIDENCE INTERVALS:"
730 PRINT "POP. FROM ";XM-C1;" TO ";XM+C1
740 PRINT "MEAN FROM ";XM-M1;" TO ";XM+M1:CHR$(17)
750 PRINT "99% CONFIDENCE INTERVALS:"
760 PRINT "POP. FROM ";XM-C2;" TO ";XM+C2
770 PRINT "MEAN FROM ";XM-M2;" TO ";XM+M2:CHR$(17)
900 REM ENDING?
910 PRINT:PRINT,CHR$(154) "ANOTHER GO? Y OR N"
920 GET G$:IF G$<>"Y" AND G$<>"N" THEN 920
930 IF G$="Y" THEN RUN
940 PRINT CHR$(147) CHR$(154) "BYE FOR NOW":END

READY.
```

Summary
(by chapter)

1. Simple functions
Displaying numbers neatly
Right-justified numbers
INT, the integral part of a number
ABS, the absolute value of a number
Rounding off numbers
Rounding up and down
Rounding off a number to D decimal places
Bank balances
Overdrawn bank balances
SGN, the sign or signum function
Displaying bank balances colourfully

2. Trigonometry
Scale drawings
Estimating heights and distances
Right angled triangles
Trigonometric functions, TAN, SIN and COS
Hypotenuse, opposite side, adjacent side
Radian
PI
Degrees to radians, radians to degrees
Finding lengths of a right angled triangle
Inverse functions
Arc tangent, ATN
Arc sine, ASN
Arc cosine, ACS
Pythagoras' theorem
Non right-angled triangles
Law of cosines
Law of sines
Finding angles and/or sides of a triangle
Refraction
Angle of incidence, angle of refraction
Snell's law
Refractive index
Reflection
Critical angle

151

3. Earth trigonometry
The Earth
Straight lines on the Earth
Great Circles
Lines of longitude
Greenwich, England
Lines of Latitude
Calculating the distance between two points on the Earth

4. Powers
Squares of numbers
Powers of numbers
Properties of powers
SQR, Square root
Imaginary numbers
Complex numbers
Complex numbers on the Commodore 64
Quadratic equations
Solving quadratic equations
Roots, finding roots
Formula for the roots of a quadratic equation
Discriminant
Solving quadratic equations on the 64
Solving other equations
Polynomial equation
Degree of a polynomial
Roots of polynomials
Newton's method
Derivative of a polynomial
Finding roots via Newton's method
EXP, the exponential function
E, EXP(1)
Factorial
Properties of the exponential function
Formula for exponential function
Logarithmic function
LOG, the natural logarithmic function
Properties of the logarithmic function
Finding roots of other functions

5. Sequences
Sequences
Terms of a sequence
Generating sequences
Arithmetic sequence, arithmetic progression

Summary

Common difference
Generating arithmetic sequences
Which would you prefer?
Geometric sequence, geometric progression
Common ratio
Generating geometric sequences
Interest, compound interest
Daily interest
Double or quit gambling
Fibonacci sequences
Generating Fibonacci sequences

6. Number bases
Decimal system
Digits
Decimal representation
Base
Coefficients of a number N to a base B
Binary number system
Hexa-decimal numbers
Converting numbers from one base to another
Numbers on the Commodore 64
PEEKing numbers on the 64
Binary form of numbers between 0 and 1
Displaying binary form of numbers between 0 and 1
Floating points
Binary exponent
Binary mantissa
$1 * (1 + 2\uparrow -24) = 1 + 2\uparrow -25$ according to the 64

7. Days and weeks
Days of the week
Zeller's congruence
Finding the day of the week for any given date
Displaying a monthly calendar for any month, any year
Date management
Pseudo-Julian date
Listing dates a specified number of days apart

8. Greatest common divisor
Common divisor, common factor
Greatest common divisor, highest common factor
Euclidean algorithm

Calculating the greatest common divisor
Least common multiple
Calculating the least common multiple

9. Primes
Prime number
Composite number
There are infinitely many prime numbers
Sieve of Erastosthenes
Prime testing
Finding factors
Large primes
Mersenne numbers, Mersenne primes
Largest known prime
Probabilistic primality testing
Fermat's little theorem
Pseudoprime to a base
Most pseudoprimes are genuine primes

10. Odds and ends
Pythagorean triplets
Primitive Pythagorean triplets
Generating primitive Pythagorean triplets
Multi-precision powers
Calculating products of large numbers accurately on the 64
Calculating arbitrary large powers accurately

11. Matrices
Matrices, rectangular arrays of numbers
M by N matrix
Square matrix
Adding matrices
Why add matrices?
Matrix multiplication
Why multiply matrices
Zero matrix
Identity matrix
Inverse of a matrix, reciprocal of a matrix
Calculating the inverse of a matrix
Simultaneous matrices
Solving simultaneous matrices

12. Codes
Cryptography
Ciphers

Substitution codes
Sample substitution code program
Matrix codes
Using matrices to cipher messages
Sample matrix code program
Public-key codes
Prime numbers and secure codes

13. Random!
Heads and tails
Tossing coins
Random numbers
Simulating coin spinning on the Commodore 64
Of dice and men
Die rolling
Probability
Simulating die rolling
Simulating two dice rolling
Playing cards
Simulating card picking
Shuffling a pack of cards randomly
Non-equally likely events
Bucket with 100 coloured buttons
Simulating button picking from a bucket

14. Meaningful data
Entering numerical data into the Commodore 64
Mean, average
Calculating the mean
Maximum and minimum
Spread, range
Calculating the max, min and spread
Standard deviation
Variance
Calculating the standard deviation
Confidence intervals
Normal distribution
Calculating confidence intervals for a population
Calculating confidence intervals for the mean
Student's t distribution

Other titles from Sunshine

SPECTRUM BOOKS

Master your ZX Microdrive
A Pennell £6.95
ISBN 0 946408 19 X

The Working Spectrum
D Lawrence £5.95
ISBN 0 946408 00 9

Spectrum Adventures
A guide to playing and writing adventures
T Bridge & R Carnell £5.95
ISBN 0 944408 07 6

Spectrum Machine Code Applications
D Laine £6.95
ISBN 0 946408 17 3

COMMODORE 64 BOOKS

The Working Commodore 64
D Lawrence £5.95
ISBN 0 946408 02 5

Commodore 64 Machine Code Master
D Lawrence & M England £6.95
ISBN 0 946408 05 X

Commodore 64 Adventures
M Grace £5.95
ISBN 0 946408 11 4

Business Applications for the Commodore 64
J Hall £5.95
ISBN 0 946408 12 2

Graphic Art for the Commodore 64
B Allan £5.95
ISBN 0 946408 15 7

ELECTRON BOOKS

Graphic Art for the Electron
B Allan £5.95
ISBN 0 946408 20 3

Programming for Education on the Electron Computer
J Scriven & P Hall £5.95
ISBN 0 946408 21 1

BBC COMPUTER BOOKS

Functional Forth for the BBC computer
B Allan £5.95
ISBN 0 946408 04 1

Programming for Education on the BBC computer
J Scriven & P Hall £5.95
ISBN 0 946408 10 6

Graphic Art on the BBC computer
B Allan £5.95
ISBN 0 946408 08 4

DIY Robotics and Sensors for the BBC computer
J Billingsley £6.95
ISBN 0 946408 13 0

DRAGON BOOKS

The Working Dragon
D Lawrence £5.95
ISBN 0 946408 01 7

Dragon 32 Gamesmaster
K & S Brain £5.95
ISBN 0 946408 03 3

The Dragon Trainer
A handbook for beginners
B Lloyd £5.95
ISBN 0 946408 09 2

Advanced Sound & Graphics for the Dragon
K & S Brain £5.95
ISBN 0 946408 06 8

Sunshine also publishes

POPULAR COMPUTING WEEKLY

The first weekly magazine for home computer users. Each copy contains Top 10 charts of the best-selling software and books and up-to-the-minute details of the latest games. Other features in the magazine include regular hardware and software reviews, programming hints, computer swap, adventure corner and pages of listings for the Spectrum, Dragon, BBC, VIC 20 and 64, ZX 81 and other popular micros. Only 35p a week, a year's subscription costs £19.95 (£9.98 for six months) in the UK and £37.40 (£18.70 for six months) overseas.

DRAGON USER

The monthly magazine for all users of Dragon microcomputers. Each issue contains reviews of software and peripherals, programming advice for beginners and advanced users, program listings, a technical advisory service and all the latest news related to the Dragon. A year's subscription (12 issues) costs £8.00 in the UK and £14.00 overseas.

MICRO ADVENTURER

The monthly magazine for everyone interested in Adventure games, war gaming and simulation/role playing games. Include reviews of all the latest software, lists of all the software available and programming advice. A year's subscription (12 issues) cost £10 in the UK and £16 overseas.

COMMODORE HORIZONS

The monthly magazine for all users of Commodore computers. Each issue contains reviews of software and peripherals, programming advice for beginners and advanced users, program listings, a technical advisory service and all the latest news. A year's subscription costs £10 in the UK and £16 overseas.

For further information contact:
Sunshine
12–13 Little Newport Street
London WC2R 3LD
01-437 4343